REVISED NUFFIELD CHEMISTRY
CHEMISTS IN THE WORLD

CONTENTS

Editors and contributors *ii*

Chapter 1 **Finding out about the atom** *1*

Chapter 2 **Using ideas about atoms** *16*

Chapter 3 **The way of discovery** *28*

Chapter 4 **Davy and Faraday** *38*

Chapter 5 **Energy and chemicals from coal, gas, and oil** *50*

Chapter 6 **Polymers from petroleum** *68*

Chapter 7 **Fertilizers** *78*

Chapter 8 **Ceramics and glasses** *87*

Published for the Nuffield Foundation by Longman Group Limited

General editor,
Revised Nuffield
Chemistry:
Richard Ingle

Editor of this volume:
J.A. Hunt

Contributors:
H.G. Andrew
Sir Lawrence Bragg
R.A. Faires
F. Greenaway
L.F. Haber
J.A. Hunt
J.L. Hunt
H.P.H. Oliver
M.J.W. Rogers
O.J. Walker

The editor wishes to acknowledge his indebtedness to H.P.H.Oliver, who edited the original series of Background Books on which this volume is based. He would also like to thank Dr F.Greenaway and Professor G.W.A.Fowles who read a preliminary draft and made many valuable suggestions for improving it. He would also like to express his gratitude to Dr A.J.Moulson, who acted as consultant for Chapter 8, and to Dr R.C.Poller for his help with Chapter 5.

All rights reserved. No part of this publication may be reproduced, stored in a retrieval system, or transmitted in any form or by any means — electronic, mechanical, photocopying or otherwise — without the prior permission of the copyright owner.

Cover picture:
Polymers in use.
BASF

Longman Group Limited London
Associated companies, branches, and representatives throughout the world

First published 1979
Copyright © The Nuffield Foundation, 1979

Design and art direction by Ivan and Robin Dodd

Made and printed in Great Britain by Whitstable Litho Limited Kent

Chapter 1
FINDING OUT ABOUT THE ATOM

FINDING OUT ABOUT THE ATOM

Early theories

Present-day chemistry is built on the foundations of the atomic theory. This theory was put forward over a hundred and fifty years ago by an English Quaker called John Dalton (figure 1). Atoms were not Dalton's original idea; long before his time many ancient philosophers had pondered on the nature of matter. Some Greek thinkers had even discussed whether matter was infinitely divisible, or whether it was made up of a very large number of tiny particles which could not be cut up. Those who favoured the 'particles' theory were known as atomists. Democritus, who was born in about 460 B.C., and Epicurus (341–270 B.C.) were the most famous of them.

In the seventeenth century in Europe ideas about atoms were revived, notably by Pierre Gassendi (1591–1655) in France and Robert Boyle (1627–91) in England. Sir Isaac Newton wrote in his *Principia* (1687) about the behaviour of the particles constituting a gas, and later discussed atoms in a more general way in his *Opticks*. 'It seems probable to me,' he wrote, 'that God in the beginning formed matter in solid, massy, hard, impenetrable, moveable particles, of such sizes and figures, and with such other properties, and in such proportion, as most conduced to the end for which He formed them.'

By the end of the eighteenth century many of the facts of chemistry had been discovered, but there was no satisfactory theory to account for them. William Higgins, an eccentric Irish chemist, had made an interesting contribution. He said that chemical reactions could be explained by supposing that particles of different substances combined. His work (1789) was done before that of John Dalton but nobody took much notice of what he said, perhaps because his work bore a title concerned with the phlogiston theory, then the topic of great controversy. Dalton did not read Higgins's book until after his own theory was announced, and in any case Higgins's theory did not have the scope and power of Dalton's theory.

Dalton's great achievement was to show that measurements of the masses of the elements which combine together can be used to work out the relative masses of atoms. He thus changed a vague idea into a practical method of study which led to the discovery of the formulae of compounds.

Figure 1
Portrait of John Dalton (1766–1844) who first introduced the idea of the atom into chemical explanation.
The Royal Society

Why do you believe in atoms? Is it because you have been told about them by teachers or read about them in books? Do any of the experiments you have done convince you that all matter is made up of tiny particles?

THE BEGINNINGS OF MODERN ATOMIC THEORY

John Dalton

Dalton was born in the remote village of Eaglesfield in Cumberland in 1766. His parents were Quakers and his father was a handloom weaver. It was fortunate for Dalton that even in so remote a village there were two men who were able to help him considerably in his early life — his teacher, John Fletcher, and a distant relative, Elihu Robinson, described as 'a man of education and property' who lived nearby. Between them these two men gave Dalton an education in mathematics and an interest in science which would have been hard to better anywhere at that time. Elihu Robinson had an important influence on his career. When Dalton told him that he felt he would like to take up medicine Robinson wrote to dissuade him, 'believing that thou wouldst not only shine, but be really useful in that noble work of teaching youth'. He started on 'that noble work of teaching youth' at a very early age. When he was only twelve years old he started a school of his own in Eaglesfield. It seems to have been quite a success despite the difficulty he had in keeping the other children in order, especially those who were older than he was! Later he moved to Kendal and opened a school there with his brother. It was in Kendal that he met John Gough, a blind man of great intelligence whose interest in science and skill in making scientific instruments influenced Dalton.

As a child John Dalton showed those qualities of determination which later made him a great scientist. He would puzzle over a mathematical problem for hours until finally his perseverance brought success. After he had become famous, he said that if he had succeeded where other men had failed it was because he had devoted himself more completely to his researches, and not because of any superior talent.

The universities of Oxford and Cambridge were closed to Dalton since entry was restricted to members of the Church of England. As a Quaker one of the few places of learning open to him was Manchester New College, a Presbyterian foundation. Through the influence of John Gough he was offered and accepted a teaching post at Manchester College in 1793, and it was in Manchester that he lived and worked for the rest of his life.

Dalton's first interest, stimulated by Elihu Robinson, was meteorology, and he kept this up until his death in 1844. By then he had made over 200 000 recorded observations, many of them while he was walking over the mountains of the Lake District; he was an extremely brisk walker, often exhausting his companions as he strode up and down the mountainsides. 'John! I wonder what thy legs are made of!' a footsore friend once shouted after him. Taking his barometer and his notebook, he would spend days with his friends happily combining the pleasure of walking with the more serious business of measuring the barometric pressure, humidity, and rainfall. How useful, he said, for sailors, farmers, and other people if it were possible to predict the weather. But there were other thoughts in his mind as he climbed Helvellyn or Skiddaw — the properties of the gases of the air.

The study of the air had led to the discoveries of Cavendish, Priestley, and Lavoisier in Dalton's own lifetime. Through further study, Dalton was guided towards his greatest achievement — the atomic theory. One of the problems which Dalton attacked was whether air was a mixture of gases or a single compound (plate 1). He had already joined the atomists and thought, with Newton, that all matter was made up of particles. But what were these particles like?

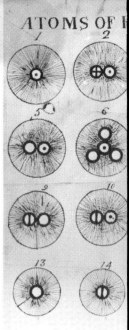

Figure 2
Diagrams drawn by Dalton to represent atoms and 'compound atoms' (molecules) of gases and liquids. The small circles are atoms of different elements and the lines radiating out from them are 'caloric' or heat.
Science Museum, London

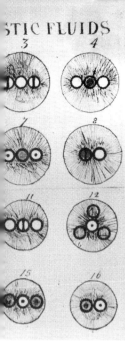

Figure 3
(Left) Dalton's list of chemical elements (1806–1807). Azote is the old name for nitrogen.
Science Museum, London

Newton had thought that the air was a single substance of identical particles. But it had since become known that air was composed of both nitrogen and oxygen, and that oxygen was heavier than nitrogen. Why then wondered Dalton, did the oxygen not sink to the bottom, unless it was chemically bound to the nitrogen? You must remember that at that time the particles were commonly thought to be stationary. The idea that they were in a state of constant motion (the kinetic, or 'particles-in-motion', theory) was not evolved until much later. Dalton drew on paper his idea of what a particle, or atom as he called it, of oxygen looked like. He drew a solid circle and round it rays like the rays of the sun to represent the elastic nature of the 'caloric' or heat which he thought surrounded the atom (figure 2). To represent the particles in the air, he tried drawing oxygen and nitrogen atoms in pairs to represent a 'compound atom' between oxygen and nitrogen.

By 1789 Lavoisier had shown that the air consisted of about 4 parts by volume of nitrogen to 1 part by volume of oxygen. Dalton assumed that the volumes of the gases had something to do with the numbers of atoms of each element that were present and, as he said himself, he 'ran out of oxygen atoms'. He therefore dismissed the idea of a compound and concluded that, in the air, the atoms of oxygen and nitrogen were not combined. The important point about this stage of Dalton's work was that he had begun to explain the behaviour of gases in terms of particles with definite individual properties. In the same spirit, he was led to suggest that all chemical reactions could be explained in terms of the combination of atoms. This was the beginning of the atomic theory.

Have you ever wondered why oxygen, which is slightly denser than nitrogen, does not sink to the bottom of the atmosphere? Can you explain why it does not do so?

Atoms and atomic weights

Dalton wrote down his ideas about atoms in his notebook. The date of his first entry was 6 September 1803. He stated that matter consisted of small particles or atoms and that atoms were indivisible and could neither be created nor destroyed: as we have seen, these ideas were not new.

He then went on to propose that the atoms of an element were identical. He emphasized, in particular, that their weights were identical and that the weight of the atoms of one element was different from the weights of the atoms of all other elements. He also proposed that the reaction between elements to form compounds could be explained by the combination of atoms. Higgins had previously suggested this, but it was Dalton who brought the idea to people's notice. It is these last two ideas — that atoms of different elements are distinguished by weight and that chemical reaction takes place through combination of atoms — that constitute Dalton's chief contribution to chemistry (figure 3). For if

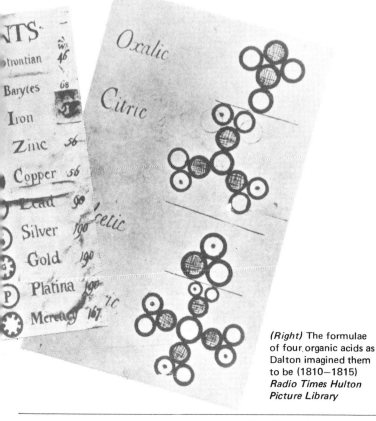

(Right) The formulae of four organic acids as Dalton imagined them to be (1810–1815)
Radio Times Hulton Picture Library

Finding out about the atom

the atomic weights of different elements were known, then the number of atoms in a given weight of any element could be worked out, and the number of atoms involved in a chemical reaction determined precisely. The problem was how to find the atomic weights of the different elements.

Obviously atoms were very much too small to weigh directly on a chemical balance and so get the *absolute* weight. Dalton suggested that it might be possible to determine the relative weights of atoms – that is, how much the atoms of one element are heavier (or lighter) than those of other elements.

Chemists immediately set about finding these atomic weights. (The term 'relative atomic mass' is now used in preference to the older term 'atomic weight'.) How they did so makes a fascinating story, but it is not one that we can cover here. At an important conference held at Karlsruhe in 1860, scientists agreed on fairly reliable estimates of the atomic weights of the elements known by then – about sixty of them.

They showed, for example, how much heavier an oxygen atom was than a hydrogen atom, or how much heavier a chlorine atom was than an oxygen atom; they did not show the absolute weight (the weight of a particular atom in millionths of a gram) of any of them.

For almost a hundred years after John Dalton put forward his atomic theory, people thought of atoms as solid, indestructible particles. They had no reason, and no experimental evidence, to think otherwise. And although we have now learned much more about atomic structure, and although our model of the atom is no longer that of a solid particle, Dalton's original idea remains a useful one. In most of the chemistry you have studied, for example, the solid atom has been a perfectly adequate basis for understanding chemical behaviour – to explain reactions in terms of rearrangements of the atoms and to arrange a periodic table of the elements based on their atomic weights. In much (but not all) of the chemistry you will do in the future, it will continue to be adequate. Dalton's model of the atom is simple. That it is too simple – that it will no longer explain everything we know about the behaviour of substances – is no reason for getting rid of it.

Look at Dalton's list of the elements, then try to devise and draw your own Dalton-style symbols to represent elements such as bromine and uranium. Which of the 'elements' in his list are not now thought to be elements?

Compare Dalton's symbols for the atoms of elements with modern chemical symbols. Which do you prefer? Why do you think that we no longer use Dalton's symbols?

OPENING UP THE ATOM

The discovery of the electron

The first step leading to the idea that atoms might not be the ultimate particles of matter, solid and indestructible, was the discovery of the electron – a 'particle' very much smaller than the atom itself. The discovery was made by scientists investigating the passage of electricity through gases.

At normal pressures, gases are poor conductors of electricity. But when an electric current is passed through a gas at low pressure, a thin zig-zag spark darts between the two electrodes (figure 4a). As the pressure is further reduced a bright column of light fills the tube (figure 4b). The colour of this light depends upon the chemical nature of the gas: the discharge of sodium gas is yellow; that of neon is red. You will have seen lighting of this kind in street lamps, in shop and advertising signs, and frequently in houses. Finally, as the pressure is even further reduced the bright column of light moves towards the anode, and the glass opposite the

Figure 4
a Discharge of electricity through a gas as the pressure is reduced. At a pressure of about 50 mm of mercury, a thin zig-zag spark passes between the two electrodes.

b At a pressure of about 10 mm of mercury, a bright column of light fills the tube.

c At a pressure of about 0.5 mm of mercury, the column of light recedes towards the anode, and the glass opposite the cathode begins to glow. An object (such as a cross) placed in front of the cathode casts a shadow on the glass.

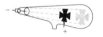

cathode begins to glow (figure 4c). You will have seen this glow too — on the screen of a television set.

In 1869 Hittorf suggested that the glow was caused by rays streaming from the cathode because he was able to show that when a solid object was placed in front of the cathode it cast a shadow on the glass.

A few years later Sir William Crookes found that these cathode rays could be deflected with a magnet. He therefore concluded that they were not light rays, as had been originally supposed, but were electrically-charged particles. Later they were found to be negatively charged.

In 1897 Sir J.J. Thomson carried out some experiments on these negatively-charged particles (now called electrons) at the Cavendish Laboratory in Cambridge (figure 5a and b). As a result of these experiments he arrived at a rough value for the mass of the electron. This was found to be very small, about two thousand times less than the mass of the smallest atom, the hydrogen atom.

- - - - - - - - - - - - - - - - - - - -

Can you say anything about the likely size of an electron now that you know that Sir J.J. Thomson found that the mass of an electron is about two thousand times less than the mass of a hydrogen atom?

- - - - - - - - - - - - - - - - - - - -

Radioactivity

While the first work on the structure of the electron was going on, another big step towards opening up the structure of the atom was made — the chance discovery of radioactivity by Henri Becquerel in 1896. In his laboratory in Paris, Henri Becquerel was fascinated by the apparent connection between X-rays and fluorescent and phosphorescent substances. Fluorescent substances emit a glow when stimulated by light. Phosphorescent substances, unlike fluorescent ones, continue to glow for some time

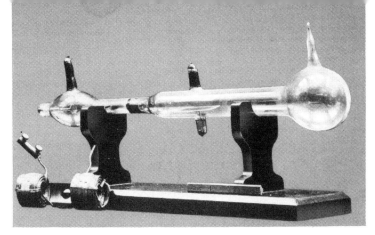

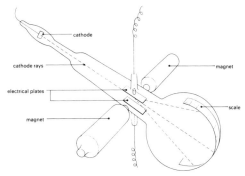

Figure 5
a Thomson's original vacuum tube in which he measured the ratio of the charge of an electron to its mass. *Lent to the Science Museum, London, by the late Lord Kelvin*
b Diagram of a vacuum tube for measuring the ratio of the charge of an electron to its mass. The cathode rays (electrons) pass along the tube through magnetic and electric fields at right-angles to each other. The deflection of the rays is observed on the scale surrounding the large bulb.

after the stimulating light source has been removed; 'luminous paint' is an example. X-rays had been discovered in 1895 by Wilhelm Röntgen, a German physicist, who was investigating fluorescence and found that even when he enclosed a discharge tube in a light-tight box, crystals of barium platino-cyanide placed nearby would glow brightly. Some kind of invisible rays from the discharge tube seemed to be escaping from the box and stimulating the crystals. Röntgen also found that the rays would pass through the walls of a light-tight box and darken a photographic plate. He gave these penetrating rays the name of X-rays, and used them to take X-ray photographs of opaque objects such as the hand. This discovery caused quite a stir: for instance 'X-ray proof clothing' was advertised in order to preserve dignity and modesty!

Now Becquerel wanted to find out whether phosphorescent uranium salts, which gave a glow similar in appearance to that of a discharge tube, also emitted X-rays. He put crystals of potassium uranyl sulphate on a photographic plate wrapped in black paper and exposed the whole to sunlight. He found that the plate was blackened in the neighbourhood of the crystals.

At first he thought that penetrating rays had been produced together with phosphorescence as a result of exposing the crystal to light; however, one dull day in 1896 he prepared an experiment of this kind but put the crystals and wrapped plate in a drawer. About five days later, he developed the plates and found that the blackening near the crystals was as intense as when he had exposed the crystals to sunlight (figure 6). He concluded that the radiations were a property of the uranium salt, quite independent of external stimulus; and so, without naming it, he had discovered radioactivity.

If you make a surprising observation in the laboratory do you ignore it or do you think about it and investigate it? Do you expect to make an important scientific discovery?

Becquerel found that the radiations from uranium penetrated materials and, like X-rays, could produce shadow-graphs. He is credited with having produced a radiograph of a key by placing it on a wrapped plate and covering it with crystals of a uranium salt. (Both this and the original experiment with uranium crystals are among the easiest experiments in radioactivity, and can be carried out in a school laboratory.)

A further property of the radiations discovered by Becquerel was that they would cause ionization in air and would discharge a gold-leaf electroscope in a similar way to Röntgen's X-rays.

Do you think that X-rays and radioactivity could have been discovered if photography had not been invented?

Figure 6
A reconstruction of Becquerel's original experiment. The photographic plate, at the bottom of the dish, became fogged — although it was covered from the light with paper. Becquerel therefore concluded that the fogging must be due to the uranium compound lying on top of the plate.
U.K.A.E.A.

The Curies and radium

Becquerel's young Polish assistant, Marya Sklodowska, later married to Pierre Curie, (plate 2) attempted to measure the ionization with apparatus which is used in various forms today, consisting of two parallel plates insulated from each other and connected through a battery to a galvanometer. When the uranium salt was placed between the plates, the galvanometer indicated a flow of current. She found that all compounds of uranium showed the phenomenon which she named 'radioactivity' and that the more uranium there was, the greater was the effect. It was found that thorium compounds were also radioactive. In the course of her investigations she observed that the uranium ore pitchblende was much more radioactive than uranium and suggested the presence of an element much more active than uranium. In fact several other minerals showed this effect. The Curies found that a highly active substance could be separated by precipitation from a solution containing uranium and thorium, together with bismuth. They were able to extract the material and show that it had far greater activity than uranium. They claimed it as a new element and suggested the name *polonium* after Marie Curie's native land. Further work showed that a second source of intense radioactivity was associated with barium, and the new element *radium* was claimed.

Figure 8
Lord Rutherford (1871–1937) and J.A. Ratcliffe.
Cavendish Laboratory, University of Cambridge.

Figure 7
The interior of the laboratory where the Curies first separated radium and polonium.
Archives Pierre et Marie Curie, Institut du Radium, Paris.

Early in the new century, the extraction of radium became an industry. The Curies freely gave information about the process of extraction and it is a tribute to their careful work that there has been very little change in the method of extraction since their time. Pierre was killed in a street accident in Paris in 1906, but Marie lived on until 1934.

The radiations from radioactive substances

Marie Curie had set out to measure radioactivity but this led on to the discovery of two new radio-elements and the start of a new industry.

In order to confirm the claim to have discovered two new elements, the Curies set to work to prepare quantities large enough to be weighed. This was a tremendous task since they started with a tonne of pitchblende residues from which uranium had been extracted. In rather primitive conditions, and turning chemists on a rather large scale, they worked for four years and were able to produce 100 mg of pure radium chloride and to establish the atomic weight of the element.

Work by a number of people established that the radiations from radium had a beneficial effect on certain growths and tumours, so, like Röntgen's X-rays, radium was added to the armoury of the medical man, and is still there, although artificial radioactive substances, obtainable in great quantities, have extended the work, and to some extent replaced radium.

The Curies observed that radium appeared to make objects near to itself radioactive whereas uranium salts did not behave in this way. Rutherford (figure 8) found that thorium salts behaved like radium, and gave off a radioactive gas which he called 'emanation' (figure 9). As the activity of the gas died away an 'active deposit' was left, where the gas had been, with its activity increased.

Rutherford and Soddy explained this by the theory that when an atom of a radioactive element has given out some radiation it becomes the atom of a new element. The new atom might itself be radioactive and it would then throw out another particle and so on. From a number of tests, Rutherford and the Curies recognized that radioactivity contained two different kinds of radiation, which they called alpha (α) and beta (β). Beta rays were found to be deflected by a magnetic field, to have a negative electric charge, and

Figure 9
Rutherford's laboratory at Cambridge.
Cavendish Laboratory, University of Cambridge.

to be identical with electrons. Consideration of the amount of bending suggested that the mass of a beta particle was rather more than $\frac{1}{2000}$ that of a hydrogen atom. It took a considerable amount of research to find out the nature of the alpha particle. It could be deflected by a powerful magnetic field but in a direction opposite to that of a beam of electrons. This showed that it had a positive charge (opposite to the charge on an electron). By considering the force needed to bend a beam of alpha particles in a magnetic and an electric field, Rutherford thought it probable that they were helium atoms which had lost two electrons.

He proved this by placing a radioactive gas (radon) in a glass tube, enclosed by another tube which he evacuated (figure 10). The walls of the inner tube were thin enough to allow the α-particles emitted by radon to pass through. After some days, the gas in the outer tube was compressed by mercury into a portion through which an electric discharge could be passed. It showed the spectrum of helium, and this was regarded as definite proof of the nature of the α-particle.

A third type of radiation was found to exist. This was not deflected by a magnetic field but had considerable penetrating power. Gamma (γ) rays, as this radiation is now called, were proved after some time to be similar to X-rays but of shorter wavelength.

Madame Curie's way of summarizing the electrical properties of the radiations was to suppose that the source of radiations was in a hole in a block of lead so that they would emerge as a narrow vertical pencil. If a magnetic field is applied perpendicular to the paper, the direction of deflection would be as shown in the diagram (figure 11). The heavier α-particles are only slightly deflected whereas the light β-particles are bent to the left to a greater extent. Gamma rays, having no electric charge, go straight on. It should be pointed out that this is a summary of these properties, and not the result of an experiment. β-particles would be bent so much more than α-particles that the difference could not be shown in the same experiment.

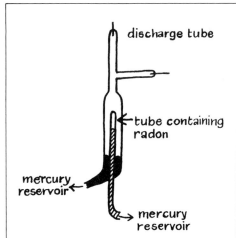

Figure 10
Diagram of Rutherford's apparatus for investigating the nature of α-particles.

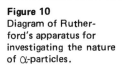

Figure 11
Effect of a magnetic field on radiations. Deflections not to scale. The dots signify the presence of a magnetic field.

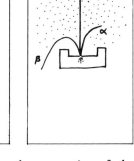

We can summarize the properties of the radiations thus:

	Mass relative to hydrogen	Charge
α	4	+2
β	1/1837	−1
γ	0	0

Look for and read a life of Madame Curie.*

ATOMS LIKE THE SOLAR SYSTEM

The discovery of elements that emitted electrically charged particles led to the idea that atoms might be built up of these particles. Scientists devised models to show what they thought the structure of the atom was like. One of the models put forward was J.J. Thomson's 'plum pudding' model in which electrons were thought to be embedded in a sphere of positive charge like currants in a bun. But it was Rutherford and his team of scientists who made the discoveries which led to the modern theories of atomic structure.

Lord Rutherford was one of the greatest of all scientific experimenters. In 1911 Marsden and Geiger, two members of the Physics Department working under Rutherford at Manchester University, carried out a famous series of investigations into the structure of

*One of the best lives is by her daughter Eve Curie, *Madame Curie, a biography*, Heinemann (1938).

the atom. They directed α-particles from a radioactive element at a target of very thin gold foil. To find out the effect of the foil atoms on the high-speed positively charged α-particles, they placed behind the foil a screen coated with phosphorescent zinc sulphide.

What this screen is and why they used it need some explaining. In all experiments aimed at probing the secrets of the atom there is the problem that individual atoms are far too small to be seen, even today with the highest-powered microscopes. However, it is often possible to reveal the presence of atoms or of atomic particles indirectly. The phosphorescent screen, when it is struck by an α-particle, produces a tiny flash of light, or scintillation, which can be seen through a microscope. You can see this effect yourself if you look at the hands of a luminous watch through a powerful lens. By observing the position of the scintillation on the phosphorescent screen, it is possible to tell through what angle an α-particle has been turned. (Another instrument for observing indirectly the effect of atomic particles is the cloud chamber (figure 12a and b).)

It was found that most of the particles passed through the gold foil with only minor deflections. This was as Rutherford had expected. A few α-particles, however, were turned through a very large angle, and occasionally an α-particle bounced back. Commenting on this, he said: 'It was quite the most remarkable event that ever happened to me in my life. It was almost as incredible as if you had fired a 15-inch shell at a piece of tissue paper and it came back and hit you.'

To account for this 'remarkable event', Rutherford formed the idea of a nucleus of concentrated positive electrical charge at the centre of each gold atom. Should a positively-charged α-particle collide directly with a positively-charged nucleus, this would explain why the α-particle was being repelled. However, because most of the α-particles passed through the foil with only minor deflections, Rutherford concluded that there must be a relatively large area of 'empty' space surrounding each nucleus.

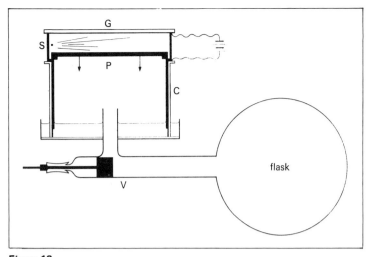

Figure 12
a Diagram of a Wilson cloud chamber — the chamber (C) is covered with a glass plate (G), and the gas in the chamber is saturated with water vapour. If the circular piston P is dropped a short distance (by opening the valve V connected to the evacuated flask), the pressure of the gas falls and it becomes supersaturated with water vapour. An α-particle from the source S ionizes the gas molecules in its path, and the supersaturated water vapour condenses on these ions, leaving a visible trail behind the α-particle.
b A fan of α-particles photographed in a cloud chamber by C.T.R. Wilson.
Crown Copyright. Science Museum, London.

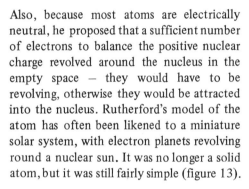

Figure 13
Rutherford's model of the atom: a nucleus of positive charge around which orbited 'planetary' electrons.
U.K.A.E.A.

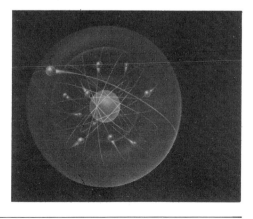

Also, because most atoms are electrically neutral, he proposed that a sufficient number of electrons to balance the positive nuclear charge revolved around the nucleus in the empty space — they would have to be revolving, otherwise they would be attracted into the nucleus. Rutherford's model of the atom has often been likened to a miniature solar system, with electron planets revolving round a nuclear sun. It was no longer a solid atom, but it was still fairly simple (figure 13).

The Rutherford model of the atom provided chemists with a new insight into the way in which substances behaved. It explained, for example, the formation of charged atoms (or 'ions' as they are called) by the loss or gain of one or more electrons. Thus, in the formation of sodium chloride, each sodium atom loses an electron to become a positively-charged ion or cation (Na^+), and each chlorine atom gains an electron to become a negatively-charged ion or anion (Cl^-).

The Rutherford model was refined by Niels Bohr to explain the pattern of the spectrum seen when light from a discharge tube containing hydrogen is viewed through a spectroscope. Bohr used the idea that the lines in the spectrum are formed when electrons jump from one orbit round the nucleus to another, giving out a definite amount of energy in the process. He was then able to work out the size of the orbits, using the novel quantum theory which had recently been devised by Max Planck to explain the pattern of radiation emitted from hot surfaces. The name quantum came from a Latin word meaning 'how much' and the theory was based on the idea that particles gain and lose energy in definite jumps and not continuously.

But within a few years it was shown that the picture of a solid electron moving in definite orbits around the nucleus could not account for all the properties of atoms.

Do you have a picture in your mind of what atoms are like? If so, does your picture resemble the Dalton atom or is it more like the Rutherford model?

WAVES AND PARTICLES

In 1897 the German physicist Heinrich Hertz had shown that if solids (especially metals) were irradiated with ultra-violet light they emitted electrons — the light was 'knocking' the electrons out of the metals. This effect could not be explained on the basis that light consisted solely of waves, as had been originally supposed. To explain this difficulty that light had the properties of both waves and particles, Albert Einstein, like Niels Bohr, used Max Planck's quantum theory and suggested that light consists of packets of energy which he called photons.

Later the reverse argument came to be applied to the electron. Summing up investigations into the behaviour of electrons, Sir James Jeans stated: 'The hard sphere has always a definite position in space; the electron apparently has not. A hard sphere takes up a very definite amount of room; an electron — well, it is probably as meaningless to discuss how much room an electron takes up as it is to discuss how much room a fear, an anxiety, or an uncertainty takes up.'

So Prince Louis-Victor de Broglie (figure 14) put forward the idea that electrons had the properties of waves as well as those of particles. De Broglie wondered why it was that the mathematics used to describe the motion of electrons in an atom involved whole numbers when the only other phenomena in physics involving whole numbers in a similar way were those of interference and vibration — properties of waves. A consideration of such problems led him, in 1923, to the conviction that matter, like light, should be thought of in terms of waves as well as of particles. Only in this way was it possible to arrive at a single theory that allowed the simultaneous interpretation of the properties of light and those of matter.

Figure 14
Prince Louis-Victor de Broglie (centre) shown here with G.P. Thomson (left), son of Sir J.J. Thomson, the discoverer of the electron, and G.F. Davisson (right). De Broglie was awarded the Nobel Prize for Physics in 1929.

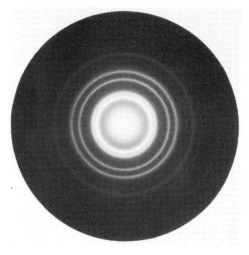

Figure 15
Diffraction patterns produced by electrons passing through a very thin film of gold. Photographs like this provided experimental evidence that electrons have wave, as well as particle, properties.
Science Museum, London. By courtesy of Professor Sir George Thomson, F.R.S.

Figure 16
H.G.J. Moseley (1887–1915) showed that the properties of a chemical element are determined largely by its atomic number or number of units of positive charge in its atomic nucleus.
The Royal Society

Whereas the wave/particle theory of light had resulted from experiment, de Broglie had conceived his theory through the 'spirit of intuition'. Because there was no experimental evidence to support it, all the leading scientists except Einstein dismissed it. If de Broglie could show that electrons produced diffraction effects, that would be sufficient. At that time, an experiment of this nature was difficult to attempt and de Broglie was not a good experimentalist.

———————————————

How many examples can you find in this chapter of an experimental discovery leading to a change in theory, or of thinking about theory leading to experimental discoveries?

———————————————

However, in 1927 Davisson and Germer in the United States and G.P. Thomson in Britain passed beams of electrons through crystals and obtained diffraction effects which are properties of waves. 'Thereafter,' said de Broglie, 'it was no longer possible to imagine an electron simply as a minute particle of electricity: a wave had to be associated with it. And this wave was not just a fiction: its length could be measured and its interference effects calculated in advance.' As a result, the simple picture of the atom with electron particles orbiting round the nucleus had to be radically altered (figure 15).

Elaborating on de Broglie's mathematical work, the Austrian scientist Schrödinger developed a system of equations known as wave mechanics. These equations could be applied to electrons irrespective of whether they were regarded as waves or as particles. At a mathematical level, the wave/particle contradiction was solved. 'At a physical level,' says de Broglie, 'the reason why these two aspects exist and the manner in which it might be possible to merge them in one superior unity, remains a mystery.' Nevertheless, he believes that one day the mystery will be solved, although, as he says, 'it will need fresh young minds to do it'.

———————————————

Which of the properties of light (and electrons) which you know about are explained using a particle theory and which are explained using a wave theory?

———————————————

THE ATOMIC NUCLEUS

Atomic number

In Rutherford's model of the atom (and, indeed, in our present-day model), the mass of the atom, and all the positive charge, are concentrated in the nucleus — the mass of the electrons is almost negligible.

In 1913 a young English physicist called Moseley (figure 16) was investigating the

X-ray spectra of certain elements. From his results, he found that the atomic nucleus of each element had a characteristic positive charge. He called this its atomic number. Thus hydrogen had a positive charge of 1 (and therefore 1 electron to balance it), helium had a positive charge of 2 (and therefore 2 electrons), lithium had a positive charge of 3 (and therefore 3 electrons). He also found that, if the elements were arranged in order of their atomic numbers, their sequence was *almost* identical with that when they were arranged in order of their relative atomic masses. It was *almost* identical but not quite — a few elements did not fit. These few elements had been the misfits in Mendeleev's original Periodic Table. Moseley's work showed that a Periodic Table based on atomic number, instead of atomic weight, produced a more logical arrangement. Subsequently, atomic number has been found to be more fundamental than relative atomic mass in determining the chemical properties of an element.

Figure 17
Professor Harold C. Urey (born 1893).

Figure 18
Aston's mass spectrograph.
Crown copyright. Science Museum, London

Look at a modern Periodic Table (see Chapter 1 of the *Handbook for pupils*) in which the elements are arranged in order of atomic number and see how many pairs of elements you can find which would be misfits in a table in which the elements were arranged in order of their relative atomic mass. How many of the misfits in the modern table had been discovered in Mendeleev's time?

Isotopes

When a lot of people become interested in a particular area of scientific investigation, it often happens that the findings branch out in a number of directions. So it was with investigations into the structure of the atom.

As long ago as 1886, the same Sir William Crookes who discovered the electrical nature of cathode rays suggested: 'When we say the atomic weight of, for instance, calcium is 40, we really express the fact that, while the majority of atoms have an actual weight of 40, there are not a few which are represented by 39 or 41, a less number by 38 or 42, and so on.' In other words, he was saying that atoms of the same element were not identical in weight, thus contradicting the views of Dalton.

As a result of his earlier work on radioactivity, Frederick Soddy came out in 1913 even more forcibly with the same view. For atoms of the same element, therefore with the same chemical properties, but of differing relative atomic masses, he coined the word 'isotope'. This is from the Greek *isos* meaning equal and *topos* meaning place; it refers to the 'equal place' of the atoms in the Periodic Table.

For some years there was little experimental evidence to support Soddy's idea of isotopes. Then in 1919 F.W. Aston investigated the masses of individual atoms with an instrument called a mass spectrograph. This instrument works on the principle of passing charged atoms through an electric field, and it showed that Soddy's prophecy was correct. It was found that the atoms of practically all elements had isotopic forms. Thus chlorine,

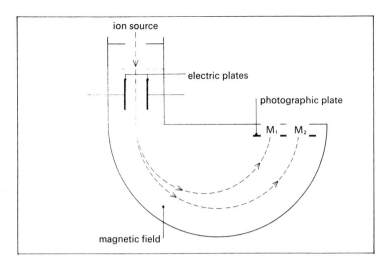

Figure 19
Diagram of mass spectrograph — a beam of positively-charged ions is deflected by a magnetic field. The extent of deflection (recorded on a photographic plate) depends on the masses of the individual ions; hence isotopes of differing mass (M_1 and M_2) show as separate regions on the plate. By using an electric field in conjunction with the magnetic field the deflection of the ions is made independent of their velocity.

with a relative atomic mass of 35.457, was shown to consist of a mixture of atoms of relative masses 35 and 37. The figure 35.457 represents the *average* relative mass of many millions of atoms. The relative masses of individual atoms were found to be very nearly whole numbers — as in the example of chlorine above, not 35.457 but very nearly 35 or very nearly 37.

―――――――――――――――――

To show you understand the meaning of the term 'isotope' try to write a definition of the word. You can read about some of the uses of isotopes in the next chapter.

―――――――――――――――――

The discovery of heavy hydrogen

During the early 1920s, H.C. Urey (figure 17) began to do research on isotopes. Little was known about them at this time, although at Cambridge F.W. Aston was investigating isotopes in his mass spectrometer. Although Aston had discovered many others, he had found no isotopes of either hydrogen or oxygen. In 1929 two isotopes of oxygen (^{17}O) and (^{18}O) were identified. This had interesting implications: according to measurements made by Aston in the mass spectrometer (figure 18), the atomic mass of hydrogen relative to that of oxygen was always in the precise ratio of 1 to 16: if some of the oxygen had an atomic mass of 18, it followed that there must be an isotope of hydrogen heavier than 1 to maintain the ratio. 'I determined', said Urey, 'to try and discover this heavier isotope of hydrogen' (figure 19). It had been estimated that, assuming the hydrogen isotope had a relative mass of 2, 1 part should be present in about 4500 parts of hydrogen of relative mass 1. Up to then it had been impossible to detect an isotope which was present in such a small quantity. Urey's chief problem was to concentrate it. He and his research assistant worked day and night to find a way of doing this. 'The only way to solve a problem of this kind is to saturate yourself in it.'

After doing pages of theoretical calculations, Urey came to the conclusion that it should be possible to concentrate the isotope by distilling hydrogen near the triple point where it exists in all three states — solid, liquid, and gas. Urey hoped that the lighter hydrogen with a lower vapour pressure would distil over first and leave a residue in which the heavier hydrogen was concentrated. Samples were prepared: 4000 cm³ of liquid hydrogen were distilled near the triple point until only about 1 cm³ of residue was left. In the autumn of 1931, Urey examined the atomic spectra of these residues under a large diffraction grating. Lines indicating heavy hydrogen were plain to see. Some of these lines, Urey subsequently discovered, were fairly visible in natural hydrogen, but it needed the concentrated samples to establish their presence.

Urey then set about investigating the properties of heavy hydrogen — or 'deuterium' as it came to be called (from the Greek *deuteros* meaning second). As suspected, it had a relative atomic mass of 2. Another hydrogen isotope, tritium, with a relative mass of 3, was discovered in 1935. Deuterium has proved a useful tracer element in biological experiments. Its nuclei (deuterons) have been extensively used in transmutation experiments. However, its main applications have been to atomic energy; it is the chief element used in fusion reactions, at present confined to the hydrogen bomb. Its oxide, heavy water, is used in large quantities to slow down neutrons in atomic reactors.

Behind Urey's discovery of heavy hydrogen there lay a piece of irony which soon came to light. Aston's measurements of the relative atomic masses of hydrogen and oxygen, from which the existence of heavy hydrogen had been predicted, were found to be in error (figure 20). As Urey commented: 'It was one of the few errors that Aston ever made, but, as he later jokingly remarked, he could hardly advocate that scientists should make mistakes on purpose so that other scientists might possibly benefit from them. It was a lucky break for me. Having started to look for heavy hydrogen, my future could have been very different if I hadn't found it.'

The nucleus

The idea of atomic number and the discovery of isotopes led to great simplifications in the interpretation of atomic structure. But, oddly enough, the kind of interpretation arrived at had been foreseen over a hundred years previously by an English physician called Prout. Prout had put forward the idea that the atoms of all elements were built up from hydrogen atoms, but his idea was eventually rejected because some atomic weights were found to contain fractions. We now know, following the discovery of isotopes, that this objection was not valid.

For practical purposes, we may regard the hydrogen nucleus, called a proton, as having a relative atomic mass of 1. Thus, since isotopes show that the relative atomic masses of all atoms are very nearly whole numbers, we could perhaps think of the nuclei of all atoms as being built up of protons. But each proton carries a single positive charge, and if atoms were built up solely of protons, their atomic numbers would be the same as their relative atomic masses.

This is not so; for example the relative atomic mass of oxygen is 16 but it has an atomic number of 8 (or 8 positive charges). What makes up the difference? This problem was resolved with the discovery by Sir James Chadwick, in 1932, at the Cambridge Cavendish Laboratory, of a nuclear particle with a mass similar to that of the proton but with no electrical charge (figure 21). It was named the neutron. At last, it seemed, the structure of the atomic nucleus was clear: it was built of positively-charged protons and electrically-neutral neutrons. Thus the oxygen nucleus consists of 8 protons (atomic number 8) plus eight neutrons so that the total relative atomic mass is 16. Or the gold nucleus consists of 79 protons (atomic number 79) plus 118 neutrons giving a total relative atomic mass of 197. The structure also explained isotopes. Thus all chlorine nuclei have 17 protons (atomic number 17) but some have 18 neutrons (relative atomic mass 35) and others have 20 neutrons (relative atomic mass 37).

As a result of further investigations we now believe this picture of the nucleus to be oversimplified. In their search to explain nuclear properties, scientists have come across many other nuclear particles in addition to the proton and the neutron. But the picture is complicated, and we cannot go into more detail here.

Figure 20
F.W. Aston (1877–1945), whose mass spectrograph enabled the masses of individual atoms to be investigated and so led to the experimental discovery of isotopes.
Cavendish Laboratory, University of Cambridge.

You can read a summary of the ideas mentioned in this section in chapter 3 of the *Handbook for pupils*.

SOME ATOMIC MODELS

As we have seen, many famous scientists have contributed to our understanding of the atom. Some have been physicists, others chemists, because here, in finding out the structure of the atom, both physicists and chemists have a mutual interest. Sometimes the discoveries have come about through planned investigation; for example, the making of new elements by the Americans. Sometimes they have come about by chance; for example, Becquerel's discovery of radioactivity and the discovery of uranium fission. Sometimes they have been predicted beforehand; for example, Soddy's prophecy about isotopes and Einstein's equation predicting the conversion of mass into energy.

It is, of course, only those discoveries and ideas which have ultimately proved useful that we hear about. Much work was done that, in the end, led nowhere. At the time it is often difficult to assess the value of a discovery or an idea. It is only on looking back that we are able to judge properly and to link up the discoveries into the coherent pattern that we have here. As a pattern it is far from complete, and many of our best scientists are still working hard to reveal more of the mysteries of the atom.

In this brief survey we have seen that ideas about the atom have changed many times. But, essentially, these ideas can be reduced to five basic models:

Dalton's solid atom,

Rutherford's 'solar-system' atom,

the atom of the late 1920s in which the wave properties of the electron were incorporated,

the atom of the early 1930s in which the nucleus was built up of protons and neutrons,

and the present-day model in which the nucleus contains, as well as neutrons and protons, many other kinds of particle.

As to future models, who knows?

These models represent the development of many ideas, each new model being a refinement of the one before, usually more complicated but able to explain a wider range of experimental phenomena (figure 22). As pointed out in the beginning, we do not necessarily drop the old model in favour of the new. In our study of chemistry, we use the simplest model that is able to help us with the kind of problem we are tackling. Sometimes this may be Dalton's model. At other times, it may be the present-day model.

Which of the experiments that you have done during your study of science can be explained using Dalton's model of the atom? Which experiments need a more complicated theory for their explanation? Which model of the atom is used in Chapter 3 of the *Handbook for pupils* to explain how atoms bond together?

Figure 21
Sir James Chadwick (1891–1974), who discovered the neutron.
Reproduced by courtesy of Lady Chadwick.

Figure 22
Atomic models: 1804, 1913, 1924, 1932, 1979, and in future?

Some atomic models

1804 Dalton's solid atom.

1913 The Bohr-Rutherford 'solar-system' atom in which electron 'planets' orbit round a nuclear 'sun'.

1924 The de Broglie atom in which the electron is no longer treated as a particle.

1932 The atom in which the nucleus is built up from neutrons as well as protons.

1979 The present-day atom in which the nucleus is built up from many kinds of particles.

Finding out about the atom   15

Chapter 2
USING IDEAS ABOUT ATOMS

In the previous chapter you have followed the development of the modern theory of atomic structure. In this chapter, you will see how an understanding of the structure of the atom has been used to create new elements, to release large amounts of energy, and to manufacture radioactive chemicals.

CHANGING ONE ELEMENT INTO ANOTHER

The alchemists dreamed of transmuting 'base' metals into gold. They failed. But, with a new insight into the structure of the atom, transmutations of this kind eventually became possible.

Radioactivity is itself a kind of natural transmutation in which the radioactive elements lose protons and electrons to become other elements. The first artificial transmutation was carried out by Lord Rutherford in 1919.

Figure 23
Cloud chamber photograph showing the transmutation of a nitrogen nucleus into an oxygen nucleus by collision with an α-particle. The oxygen nucleus makes the short track almost in line with the colliding α-particle. The long track at about 110° to the α-particle track is a proton, the other product of the nuclear reaction.
Science Museum, London. By courtesy of Professor P.M.S. Blackett, F.R.S.

Using a radioactive source, he bombarded nitrogen gas with α-particles (helium nuclei) and, by dislodging a proton, managed to convert a few of the nitrogen atoms into oxygen atoms (figure 23).

How can we represent transmutations of this kind in the form of an equation? In representing the transmutation of one element into another, we are concerned not so much with the number of atoms (as in conventional equations) as with the nuclear structure of the individual atoms. We want to know how many protons each atomic nucleus contains (atomic number) and the total number of protons and neutrons (mass number). Therefore, we write the atomic mass at the top and the atomic number at the base of each chemical symbol.

For example, $^{17}_{8}O$ is an oxygen isotope containing 8 protons and 9 neutrons, and $^{4}_{2}He$ is a helium nucleus (α-particle) containing 2 protons and 2 neutrons. In presenting the transmutation, we must be sure that the number of protons and neutrons on the left of the equation balances the number of protons and neutrons on the right. Thus, in Rutherford's experiment:

$$^{14}_{7}N + {}^{4}_{2}He \rightarrow {}^{17}_{8}O + {}^{1}_{1}H$$
nitrogen   α-particle   oxygen   proton
nucleus   isotope

Subsequently, Rutherford and Chadwick carried out similar transmutations among many others of the lighter elements.

To bring about nuclear transmutations, the projectile particles must have very high energies. Rutherford used α-particles from a radioactive source, but in 1932 Sir John Cockcroft and Dr E.T.S. Walton, working in Rutherford's laboratory in Cambridge, used artificial projectiles instead of natural ones.

Rutherford, in his Presidential Address to the Royal Society on 30 November 1927, had urged 'the development of sources of atoms and electrons with an energy far transcending that of α-particles and β-particles from radioactive matter'.

When a beryllium nucleus (^9_4Be) is bombarded with an alpha particle the main conversion product is carbon ($^{12}_6C$). Can you write an equation to represent this change? What is the other product?

The higher the energy of a particle, the more likely it is that when it collides with an atom it will penetrate the potential barrier of positive charge surrounding the nucleus and bring about a nuclear transmutation. An α-particle from a natural radioactive source already has a very high energy. 'Fortunately,' Cockcroft said, when interviewed about his work, 'I was saved from trying to take up Rutherford's challenge as it stood, by the timely arrival at the laboratory of George Gamow.'

Gamow, a young physicist from Leningrad, introduced a theory according to which the energy needed by a nuclear particle to break out of an atomic nucleus was less than what was generally believed at that time. Cockcroft immediately saw the consequences of this theory. 'It occurred to me,' he said, 'looking at Gamow's theory the other way about, that the energy a particle would need to penetrate a nucleus might also be less than was believed. I did some calculations which showed that it might be possible to bring about transmutations using light projectiles such as protons.' Unlike naturally-generated α-particles, protons (which are hydrogen ions) could be produced in very large numbers.

It was open to question whether or not it would be possible to accelerate them to the energies that Cockcroft's calculations showed necessary. Work in America on accelerating electrons with an electrical source of 300 kV suggested that it might be. And this, Cockcroft believed, is often the key to scientific discovery: keeping a very close watch on unusual developments in one's own line of work, studying their implications, and following them up.

Figure 24
Sir John Cockcroft in the Cavendish Laboratory, Cambridge, at the time of his famous experiment.
Keystone Press Agency Ltd.

Figure 25
Diagram of the apparatus built by Cockcroft and Walton for bombarding lithium with protons.

Cockcroft sent a memorandum to Rutherford showing that transmutations with protons were theoretically possible. Rutherford told him to go ahead and try. In this work, Cockcroft was joined by E.T.S. Walton, an Irish research student whom Cockcroft described as a 'very brilliant experimentalist'. Their problem was to build a sufficiently powerful proton accelerator. Cockcroft, who had come to the Cavendish with a degree in electrical engineering, was fortunate in combining nimble fingers with a nimble mind (figure 24).

The apparatus that they built consisted of a vacuum tube producing beams of protons (hydrogen ions) which were discharged into a stack of glass cylinders forming the accelerator tube (figure 25). By producing a high

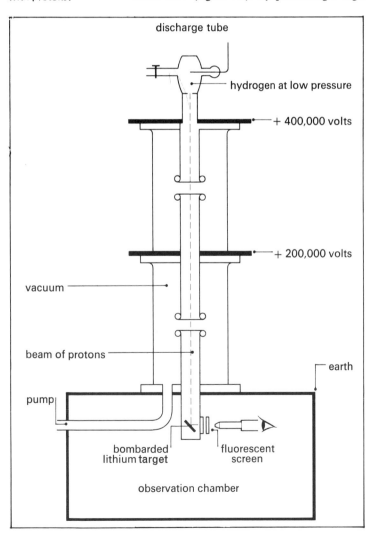

vacuum in this tube and by putting a high voltage of over 300 kV across it, they hoped to be able to accelerate protons to sufficiently high energies to penetrate the nuclei of some of the lighter elements: the lighter the element, the lower the electrical charge on the nuclei and therefore the lower the potential barrier. Their idea was to put a target of lithium, the lightest metal, in the path of the beam of protons and to observe transmutations, if any, on a scintillation screen.

They worked patiently for two years to build up and test the equipment. Their chief difficulty was to get a sufficiently good vacuum. Much of their time was spent sealing up the joints which were made with plasticine. Despite these makeshift methods, their apparatus, which cost over £500, was by far the most expensive in the laboratory. Not unnaturally, Rutherford became impatient that they should get some results from it.

On the morning of 13 April 1932 they decided to try it out on the lithium target. Flashes on the screen revealed the presence of α-particles as well as protons. These α-particles were a certain sign that protons were penetrating some of the lithium nuclei and transmuting them into α-particles (helium nuclei). For the first time, one element was being changed into another by projectiles of man's own making.

$$_3^7\text{Li} + {}_1^1\text{H} \rightarrow {}_2^4\text{He} + {}_2^4\text{He}$$

lithium nucleus   proton   two α-particles

In this transmutation of lithium with protons, there is considerable release of energy. At that time, this work represented about the best experimental test of Einstein's relationship between mass and energy ($E = mc^2$) put forward some twenty-five years before. Rutherford, on being questioned about the possibility of harnessing the energy of the atomic nucleus, dismissed it as moonshine — which, of course, at the time of the experiments it still was. But within a few years, with the fission of the uranium nucleus, the harnessing of atomic energy was to become a

Figure 26
One of the first cyclotrons — designed by Lawrence and Livingstone. Charged atomic particles enter the cyclotron through the tube, bottom left. The particles are accelerated by a series of voltage kicks along an outwardly-spiralling path, and emerge through the tube at the top as high-energy atomic projectiles.
Lent to the Science Museum, London, by the late Professor E.O. Lawrence.

Figure 27
Frederick Joliot (1900 –1958) and Irene Joliot-Curie (1897– 1956) the first to build up heavier elements from lighter elements.
E. Wehner.

reality. Making it a reality was a project with which Cockcroft himself became closely involved. He was responsible for the building of the first atomic pile outside the United States and for the postwar development of atomic energy in Britain.

─────────────────────────

In Einstein's equation E represents energy, m mass, and c the velocity of light ($c = 3 \times 10^8 \ m \ s^{-1}$). To appreciate the possibility of there being enormous energy releases in nuclear changes, use the equation to work out the energy equivalent to the mass of a proton ($m = 1.67 \times 10^{-27}$ kg).

─────────────────────────

At about the time of Cockcroft and Walton's experiments, other machines able to accelerate electrically-charged particles were coming into being. Perhaps the best known of these machines was the cyclotron, invented by the American E.O. Lawrence whose name is remembered in the man-made element lawrencium (atomic number 103). These machines are now able to produce atomic projectiles with very high energies indeed, and have led to one of the most fascinating of the recent advances in chemistry — the making of new elements (figure 26).

Building up new elements

The first transmutations involved breaking down heavier elements. Obviously the reverse process — building up lighter elements into heavier ones — soon began to interest scientists. The first scientists to achieve this (in 1934) were Irene Joliot-Curie, a daughter of Madame Curie, and her husband Frederick Joliot (figure 27). They converted some aluminium into phosphorus by bombarding the aluminium with α-particles:

$$^{27}_{13}Al + \ ^{4}_{2}He \rightarrow \ ^{30}_{15}P + \ ^{1}_{0}neutron$$

The heaviest natural element is uranium with an atomic number of 92 and an atomic mass of 238. In 1940 American chemists working at Berkeley in California began building up new elements starting from uranium.

E.M. McMillan was working at this time in the Radiation Laboratory at Berkeley. He was following up earlier experiments of the German chemist Otto Hahn. Hahn and his co-workers had been the first to consider the possibility of building up heavier elements from uranium, but their experiments in bombarding uranium with neutrons produced fission of uranium which, if it resulted in a huge release of energy, did not, as far as they could judge, lead to the formation of any heavier elements.

However, on careful examination of the fission products, McMillan identified a radioactive element with a half-life of $2\frac{3}{10}$ days — this was an isotope of element 93 (neptunium). The kinds of reaction involved in making neptunium (atomic number 93) and plutonium (atomic number 94) are:

$$^{238}_{92}U + \ ^{1}_{0}neutron \rightarrow \ ^{239}_{92}U + \gamma\text{-ray}$$

$$^{239}_{92}U \rightarrow \ ^{239}_{93}Np + \ ^{0}_{-1}electron$$

Neptunium rapidly decays as follows:

$$^{239}_{93}Np \rightarrow \ ^{239}_{94}Pu + \ ^{0}_{-1}electron$$

McMillan was then transferred elsewhere to do research on radar. Seaborg took up the research where McMillan left off, to produce and identify element 94.

─────────────────────────

The basic principles of radioactivity and the meaning of the term 'half-life' are explained in Chapter 8 of the *Handbook for pupils*.

─────────────────────────

Using the large cyclotron in the Radiation Laboratory, Seaborg (figure 28) and his co-workers accelerated deuterons (heavy hydrogen nuclei) to bombard a beryllium target which, in turn, produced fast neutrons. These neutrons were slowed down by paraffin

Figure 28
Glenn Seaborg (born 1912) in the laboratory where much of the early work on making plutonium was done.
Lawrence Radiation Laboratory, University of California.

which surrounded about 3 kg of uranium nitrate. Although the bombardment procedure may be complex, the chief difficulty in this kind of work is to separate and identify the very small quantity of the new radioactive element that is produced. Seaborg separated element 94 from the residual uranium by solvent extraction in water and ether; the uranium dissolved in the ether, leaving element 94 ('plutonium', as he called it, after the planet Pluto) in the water.

He had suspected that one of the isotopes of plutonium might be a fissionable material like uranium-235 and he produced half a microgram specially to test this. The test proved his suspicion correct. Nowadays, plutonium produced in reactors has become an important atomic energy fuel. Subsequently Seaborg detected minute traces of natural plutonium (less than one part per million million) in mineral ores such as pitchblende and carnotite.

The next step in the research programme — making elements 95 and 96 — proved to be very much more difficult. After two years' work, Seaborg was still unable to identify elements 95 and 96 among the reaction products. His lack of success merely served to make him work harder and think more deeply about the problem. Then he had a brainwave: it occurred to him that these heavy elements, starting with actinium (element 89), formed a separate series in the Periodic Table analogous to the rare-earth series (elements 57–71). Consequently, the isolation of elements 95 and 96 needed separation techniques different from those that had been used for neptunium and plutonium. Using different separation techniques, element 96 (curium) was isolated almost overnight and element 95 (americium) a few months later.

Despite this success, many people remained sceptical about Seaborg's idea that these elements formed a separate series and it was some years before the idea was generally accepted. As Seaborg says, 'Most ideas go through two stages: at first people think you're crazy; then a long time afterwards, they think you were slow not to have thought of it before.'

Following the analogy between the heavy elements and their rare-earth counterparts, elements 97 to 103 have been made and identified. The quantities produced have tended to become smaller and smaller,

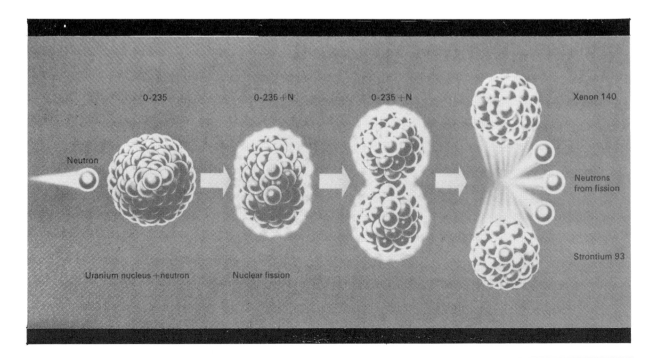

sometimes only a few atoms, and their half-lives have tended to be shorter and shorter, sometimes less than an hour. It has become progressively more difficult to identify them. With the synthesis of element 103 (called 'lawrencium', after the inventor of the cyclotron) in 1961, this series of the Periodic Table is now complete (see Chapter 1 of the *Handbook for pupils*).

How did the new elements get their names? You might list the names of the elements with atomic numbers greater than 92 and then attempt to find out after whom or where or what they are named.

ENERGY FROM THE ATOM

In Cockcroft and Walton's experiment, mass was converted into energy, but at the time, the changing of a few lithium atoms into a few helium atoms provided little insight into what was in store. How atomic energy could be harnessed on a large scale was discovered almost by chance when Otto Hahn and his colleagues in Germany found that the uranium nucleus split in half to form two very much lighter elements (figure 29).

This conversion involved a substantial loss of mass and hence a large release of energy; but, even more significant, the uranium nucleus, when it broke down, released more neutrons. These neutrons, in turn, were able to break down more uranium nuclei, and so on, leading to a chain-reaction with an immense release of energy.

It was subsequently discovered that it was not uranium with a relative atomic mass of 238 that underwent this fission reaction but a much rarer isotope with a relative atomic mass of 235. However, uranium-238 can be converted fairly readily into plutonium, as McMillan found, and this element behaves in much the same way as uranium-235. You may have heard of a 'breeder' reactor: it is in a reactor of this kind that the conversion from uranium-238 to plutonium is made.

Figure 29
Diagram showing the fission of a uranium nucleus when struck by a neutron. Note that it is the nucleus (without electrons) and not the atom which is shown. Note also that the fission products include more neutrons, thereby generating a chain reaction.
U.K.A.E.A.

These, then, are the principal atomic fuels – uranium-235 and plutonium.

As you will know, atomic energy can be used in two very different ways. In complicated equipment called an atomic pile, the fission chain-reaction is carefully controlled by slowing it down, and the nuclear energy can be 'tapped' from the pile and converted into electricity or other useful energy forms. In an atomic bomb, the fission reaction proceeds in an uncontrolled manner, and the destructive energy release is enormous. Also, the uranium nuclei break down into highly radioactive elements which greatly add to the devastation.

The fact that atomic energy came into prominence at the start of the war meant that the first efforts of scientists were devoted to making use of its destructive power. The story began with a letter from Albert Einstein to Franklin D. Roosevelt, the American President, in August 1939:

'In the course of the last four months it has been made probable through the work of Joliot in France as well as Fermi and Szilard in America – that it may become possible to set up a nuclear chain-reaction in a large mass of uranium by which vast amounts of power and large quantities of new radium-like elements would be generated. Now it appears almost certain that this could be achieved in the immediate future.

'This new phenomenon would also lead to the construction of bombs, and it is conceivable – though much less certain – that extremely powerful bombs of a new type may thus be constructed. A single bomb of this type, carried by boat and exploded in a port, might very well destroy the whole port together with some of the surrounding territory. However, such bombs might very well prove to be too heavy for transportation by air.'

The first nuclear chain-reacting pile, built by the brilliant Italian scientist Enrico Fermi in a Chicago squash court, began working in 1942. The story culminated in the dropping of two atomic bombs on the Japanese cities

of Hiroshima and Nagasaki in August 1945 (figure 30).

After the war people confidently expected that atomic fission would bring to the world vast quantities of energy at practically no cost. This dream has not come true. But, although it has so far been less easy to harness the energy of the atom than had been expected, power stations using atomic fuel are in operation in many parts of the world (figure 31) and it is certain that atomic energy will be increasingly used in the future (plate 3).

You can read more about the story of the development of the atomic bomb in *Brighter than a Thousand Suns* by R. Jungk, published in 1970 by Penguin.

Have you heard of the term 'the plutonium economy'? Do you understand why many people fear that we risk disaster if we become dependent on nuclear energy? What are the hazards associated with the use of nuclear fuels?

One other source of this energy that we have not yet mentioned is fusion. Two nuclei of deuterium (an isotope of hydrogen containing

Figure 30
Sketch of Fermi's nuclear reactor built in a Chicago squash court. The uranium fuel and the graphite for moderating the chain reaction were built up into a great pile, since when the nuclear reactors have been commonly called atomic piles.
U.S. Army Signal Corps

Figure 31
Calder Hall, Britain's first nuclear power station, and the first in the world to produce electricity on a full commercial scale, was opened by H.M. The Queen in October 1956.
U.K.A.E.A.

a neutron as well as a proton) can, in extreme temperature conditions, be made to fuse to form a helium atom. In this fusion reaction, the loss of mass and the energy of release are very much greater than in the conventional fission reaction — as witness the Sun, which produces its energy in this way. You may wonder how it is that with some atoms fusion results in a loss of mass, while with other atoms the opposite process of fission results in a similar loss.

Generally, loss of mass results from fusion between lighter nuclei (that is, with a relative atomic mass less than about 40 — for example, hydrogen) and from fission of heavier nuclei (with a relative atomic mass greater than 40 — for example, uranium). The explanation of this is not simple.

So far it has proved impossible to control the fusion reaction, and its use has been restricted to the hydrogen bomb. Scientists are trying to control the process by bottling up the very hot, ionized gas required for the fusion reaction in magnetic fields (figure 32). In October 1977 the EEC agreed to site the Joint European Torus Experiment (JET), the Community's nuclear fusion project, at Culham in Oxfordshire.

USING RADIOACTIVE CHEMICALS

Artificial radio-isotopes

Before the Second World War virtually the only radioactive material in use was a little radium used in hospitals. Quantities of radioactive substances are measured in curies where 1 curie is the quantity of radioactive material which undergoes 3.7×10^{10} radioactive transformations per second. Now many millions of curies of activity have been produced artificially, and one source used commercially may be equivalent to about 300 kg of radium.

Although there are only 90 natural elements, there are over 700 artificial radioactive isotopes. This means that each element may have a number of radioactive isotopes, some produced by bombardment of its own stable isotopes and others by different bombarding processes. Some of the radioactive products have such short half-lives that there is scarcely time to measure them by decay, and special methods are necessary. Sodium has only one stable isotope. If we add a neutron to it we get ^{24}Na with a half-life of fifteen hours. If we bombard magnesium with neutrons we get some ^{24}Na and a little ^{25}Na and ^{26}Na, but the half-lives of the last two are measured in seconds. If magnesium is bombarded with deuterons, that is, the nuclei of heavy hydrogen, consisting of one proton and one neutron, the result is ^{22}Na with a half-life of 2.6 years. We get other isotopes of sodium, ^{20}Na and ^{21}Na, by bombarding neon with protons, but they have short half-lives.

It is simple to look at the situation with the aid of nuclear equations. We cannot destroy mass, although we can exchange it for energy. We cannot destroy charge, but only redistribute it. The sodium nucleus has a charge of +11 because it contains eleven protons. A neutron has no charge and a mass of 1, a proton has charge 1 and mass 1, a deuteron has charge 1 and mass 2.

^{24}Na: $\quad ^{23}_{11}\text{Na} + ^{1}_{0}\text{n} \rightarrow ^{24}_{11}\text{Na}$

    radioactive sodium with one extra neutron

^{22}Na: $\quad ^{24}_{12}\text{Mg} + ^{2}_{1}\text{H} \rightarrow ^{22}_{11}\text{Na} + ^{4}_{2}\text{He}$

    magnesium + deuteron = sodium + an alpha particle

Figure 32
The DITE (Divertor Injection Tokamak Experiment) during final stages of construction.
U.K.A.E.A.

Figure 33
The Amersham cyclotron vault.
The Radiochemical Centre, Ltd, Amersham

Many of these isotopes are produced in Great Britain by the Radiochemical Centre at Amersham (figures 33 and 34). A simple chemical is produced or the radioactive material is left as the element, but very often the final result is a compound containing radioactively labelled atoms. For instance, $H_2{}^{35}SO_4$ or $Na^{131}I$. ^{14}C is made a constituent of a wide range of organic compounds such as drugs, and by this means it is possible to trace how they are distributed in the body.

Using ideas about atoms

Some artificial radio-isotopes				
	half-life		radiation	produced from
^{3}H	12.4	years	β only	^{6}Li
^{14}C	5770	years	β only	^{14}N
^{24}Na	15	hours	β and γ	^{23}Na
^{33}P	14.2	days	β only	^{31}P or ^{32}S
^{35}S	87.1	days	β only	^{34}S or ^{35}Cl
^{60}Co	5.25	years	β and γ	^{59}Co
^{110}Ag	253	days	β and γ	^{109}Ag
^{131}I	8.04	days	β and γ	^{130}Te
^{192}Ir	74	days	β and γ	^{191}Ir
^{198}Au	2.7	days	β and γ	^{197}Au

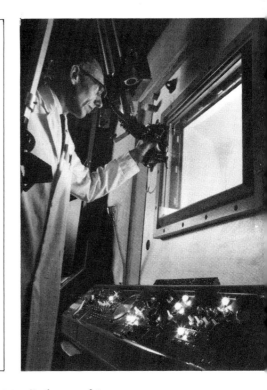

There is a large family of compounds labelled with ^{3}H (or tritium) and others labelled with ^{35}S or ^{32}P and so on.

Tracers

Since radioactivity cannot be destroyed, and since it is possible to detect very small amounts of radioactive material by reason of the radiations they emit, radio-isotopes can be ideal tracers.

Suppose we want to find out how a piston ring in an engine wears. We could weigh it, use it for a few hundred hours and weigh it again, but this would involve stopping the engine and it would be difficult to repeat the conditions of testing. If the piston ring is made radioactive, however, the small particles which wear off will be radioactive and will go into the oil system.

If the oil is then made to pass through a detector of radioactivity, we shall have a measure of the amount of wear. It is easy enough then to see the effect of load or of varying the temperature or the oil. Since the changes in wear are shown outside the engine as they occur, and the method is sensitive to small variations, it is the most efficient.

There are important medical uses of tracers (figure 35). Sometimes people have a restriction of their blood circulation. By injecting a small amount of radiosodium, the rate of flow between points can be measured and compared. Again, in the case of a skin graft, the blood flow into the new area can be determined. By choosing the appropriate radioactive material, often as a radioactively labelled compound which has been synthesized, the functioning of many body organs may be checked (figure 36).

Radioactive labelled nutrients can help the plant physiologist to study the rate and method of uptake of fertilizers and trace elements, and he can study photosynthesis by growing plants in an atmosphere of ^{14}CO$_2$.

These are only a very few of a large number of tracer applications. They may show you that they are a powerful tool of research and technology.

— — — — — — — — — — — — —

Can you think of ways in which radioactive chemicals might be used to find out (a) whether a particular method of disposal of

Figure 34
To make artificial radio-isotopes a cyclotron irradiates a rotating, water-cooled target, which is incorporated into a remotely controlled rapid target exchange system. The operator is seen here at the control panel, looking through a vault window.
The Radiochemical Centre, Ltd, Amersham

Figure 35
An International General Electric Co. gamma camera with a patient positioned for a scintigram of the brain. The camera records the distribution of radio-isotopes within the organ, and, from comparison with normal image, a diagnosis can be determined.
The Royal Sussex County Hospital, Brighton

sewage into the sea contaminates nearby beaches, (b) whether a drug being given to a patient is carried to the organs which need treatment, and (c) how fast a liquid flows along a pipeline?

Analysis

By irradiating specimens with high-energy neutrons in a reactor it is possible to determine which elements are present and how much of each element is present. If two small masses of an element are irradiated in a reactor under exactly the same conditions, the only thing which makes their activities different is their difference in mass, even when the element is in chemical combination with other elements or mixed with other materials.

Thus if 1 mg of pure copper is irradiated and shows an activity measured as 60 000 counts per minute and an unknown specimen is irradiated and gives 36 000 counts per minute, both these measurements being made with the same detector and being proportional to the actual radioactivity present, then the mass of copper in the unknown specimen is 36 000/60 000 or 0.6 mg.

In practice, it is possible to determine copper in amounts of about 10^{-10} g, and to be quite sure that it is copper because the half-life and energy of the radiations are characteristic.

Since there are some seventy elements which can be determined in amounts which are a very small fraction of a microgram, this is clearly a powerful new tool for the analyst. An important advantage is that it is possible to be quite sure what is present, and any impurities which get in during chemical extraction or manipulation make no difference.

Sometimes it is not necessary to do any chemistry to get the answer, and in fact the irradiated material remains as it was before irradiation. This has been done to find arsenic

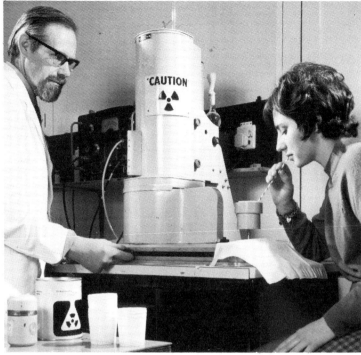

Figure 36
In the treatment of hyperthyroidism the patient is given an oral dose of iodine–131. The amount of iodine taken up by the patient's thyroid gland is measured by a scintillation counter.
U.K.A.E.A.

Figure 37
This γ-spectrometer records the radioactivity of potassium, thorium, and uranium in rocks of the sea-bed.
U.K.A.E.A.

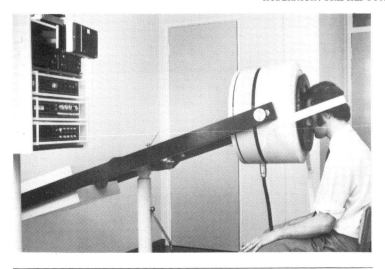

in hair or nail clippings, or to identify trace elements in dust so as to prove its origin. Thus not only does the chemical analyst have a sensitive and accurate new method, but the criminologist and forensic chemist have a new device to fight crime.

Using a reactor like 'BEPO' at Harwell it is possible to determine arsenic, gold, and many of the rare earths down to 10^{-11} g, elements such as antimony, cobalt, copper, phosphorus, and manganese down to 10^{-10} g, and most of the common elements in amounts between 10^{-9} and 10^{-7} g. The method is used for very small samples or for special investigations, and most particularly to check the accuracy of other methods like colorimetric determinations, which are simpler and do not need a reactor.

— — — — — — — — — — — — — —

Radioactive chemicals can be used to measure volumes of fluids which are otherwise difficult to determine. A small sample of the radioactive substance is thoroughly mixed with the fluid and the amount of dilution is then measured.

For example, a 1 cm³ sample of blood giving 2.2 x 10⁶ counts per minute, from a radio-isotope of long half-life, was injected into a patient's body and allowed to mix well with the blood in the body, then a 1 cm³ sample of blood was withdrawn and found to give 250 counts per minute. Can you use this information to calculate the volume of blood in the patient's body? Why was a radio-isotope with a long half-life chosen for the experiment?

— — — — — — — — — — — — — —

Gauges

How do you measure the thickness of a piece of paper? Count a large number of sheets and measure with a ruler? Cut a piece and weigh it and work it out from density and area? These methods will do, but they are not convenient if paper is being made by the mile, though the second method was used until a few years ago, and is still used for an ultimate check, and to calibrate other methods. The modern method uses β-radiation because β-radiation is absorbed by material, and the more material, the more absorption.

Therefore if a radioactive source is placed on one side of the sheet and a detector on the other, the number of β-particles which pass through is a measure of the thickness. Nothing touches the sheet, so it can be moving rapidly. Also, the signal from the detector can be continuously recorded and control the process which governs the thickness.

Often the thickness is compared with a standard, so the indication is given as the deviation from this (plate 4). Over the past ten or fifteen years, gauges have been developed and are in use in many industries for controlling the thickness of paper and sheets of plastic and metal, or the packing density of tobacco in cigarettes (which are made in a long continuous tube). None of these materials becomes radioactive because the energy of the β-particles is too small to induce radioactivity in the material being measured.

As a slight variation on this, but still in the field of gauging, consider level gauges and empty-packet detectors. The first can be used as a convenient method of indication or even of control of the level of liquid in a closed vessel. It has been used to discover how much fluid is in a fire extinguisher, or how much of a liquefied gas is present in a vacuum vessel. All that is needed is a source of radiation and a detector (figure 38a).

The same principle applies to empty-package detectors. The gauge can be set to reject an unfilled package (packet of pills, tin of paint — in fact, almost anything which is filled in an automatic plant), or it can be set to reject partially filled containers. The scheme needs no contact or weighing, can work fairly quickly, and is much used in industry.

Other uses

β- and γ-radiations are scattered by the

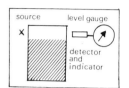

Figure 38
a Diagram of a level gauge.
b The irradiation plant at Gillette's Reading factory, with a capacity of 750 000 curies of cobalt—60. Surgical products are usually given a dose of 2.5 megarads, enough to sterilize the contents of a hermetically sealed air-tight pack until it is opened.
U.K.A.E.A.

electrons of the atoms in their path, and are sometimes sent back the way they came. The amount which comes back depends on the nature of the material which is causing the scattering. This is used to measure the thickness of plating on metal and it is the principle which guides an automatic coal cutter to remove coal and to turn aside from the rock above and below it (chapter 5, figure 66).

γ-radiation passes through metals and is reduced in intensity more or less, according to the thickness of the metal or its nature. If there were a hole or a piece of slag, this would alter the amount of radiation passing through locally. If a photographic film is put the other side, the 'radiograph' which is produced gives information about flaws, variations in thickness, and so on, just as an X-ray machine would. It takes longer to get a γ-radiograph than an X-radiograph, but very often it has a number of advantages, and among them are cheapness and simplicity.

If a large dose of γ-radiation passes through bacteria, they are killed. They need about 2.5 million rads for a fairly certain 'kill', but higher, more complex living systems are destroyed by less. Animals, including man, cannot long survive 1000 rads; weevils and cockroaches are killed by a few tens of thousands of rads. The rad is the unit of radiation dosage which is a dose of 10^{-2} J (joules) per kg of mass.

A very important use of radiation is for the sterilization of medical items — hypodermic syringes, for instance, which can be made for a few pence and are thrown away after use. This reduces the risk of cross-infection (figure 38b).

Many other items such as drugs which would be damaged by heat, sutures, scalpels, and more complicated devices are dealt with similarly.

The type of plant used commercially is very heavily shielded and is provided with safe methods for getting the materials for sterilization in and out. It may contain a few hundred thousand curies of ^{60}Co, and there are foolproof methods of ensuring that the source is safely stowed before anybody can enter. Items for sterilization are usually packed in individual sealed plastic bags, and a batch of these goes through the plant together with a 'tally' which changes colour in a significant way when it has had a sterilizing dose of radiation. Passing radiation through these materials does not make them radioactive. Many millions of items have been sterilized in this way, and it is already rare to find a doctor using the older type of hypodermic syringe, for instance, since the plastic one in its envelope labelled 'sterilized by γ-radiation' is usually preferable in every way.

Beams of γ-radiation are used for therapy instead of beams of X-rays when deep penetration is needed. Fixed sources in parts of the body can give similar radiation treatment to that given by radium. The cost is negligible, and suitable sources are very readily available for specialist hospitals.

Chapter 3

THE WAY OF DISCOVERY

Figure 40
Kekulé's Dream.

FACTS AND THEORIES

This chapter is based on the experiences of four famous scientists who have been awarded the Nobel prize (figure 39). The chapter illustrates how the interests of the scientists were first awakened and how the realization of a discovery flashed upon them. These scientists shared certain qualities: a devotion to their work and the ability to give hours of unremitting attention to it; the need to find out the answers for themselves — the *desire* to find the truth.

Frequently one realizes that the scientist

Figure 39
The Nobel Medal, which is given to the winners of the annual prizes for physics, chemistry, medicine, literature, and the promotion of peace. The awards are made from the bequest of Alfred Nobel, the Swedish inventor of dynamite, who died in 1896.
The Swedish Institute

makes use of the unexpected. He has to learn to trust, as Linus Pauling describes, the mysterious ways of the creative imagination. Science in fact depends utterly upon what science knows least about: how the human brain works; how man knows that he knows; how he can make contact within himself with the still consciousness that finds order in the moving universe.

This dependence upon faculties which are known only by their results and not by their means of working is shared by scientists with poets and composers. Let us take one example.

In 1865, the German chemist Kekulé was trying to work out the structure of benzene, which in many of its reactions behaves quite differently from most other simple organic compounds. He tells the story of how he was sitting in front of the fire and fell into a half-sleep: 'The atoms flitted before my eyes. Long rows, variously, more closely, united; all in movement wriggling and turning like snakes. And see, what was that? One of the snakes seized its own tail and the image whirled scornfully before my eyes. As though from a flash of lightning I awoke; I occupied the rest of the night working out the consequences of the hypothesis.'

So emerged the picture of benzene as a hexagonal ring (figure 40) — previously the structures of all organic compounds had been thought to be chains. The benzene ring brought many changes to organic chemistry, and it is small wonder that Kekulé was to offer this famous advice to scientists: 'Let us learn to dream, gentlemen, and we may perhaps find the truth.' However, he was then careful to add: 'But let us beware of publishing our dreams before they have been tested by a discerning mind that is wide awake.'

Many scientific discoveries are, like that of the benzene structure, of a theoretical kind. They are *explanations* of the experimental facts but are not themselves facts. Kekulé had no way of demonstrating the physical existence of a benzene ring; if it existed at all, it was far too small to be seen, even under the highest-powered microscope; but it did explain satisfactorily the known experimental facts about benzene. If facts emerged that it failed to explain, the scientists would have to think again – to date, all the facts have tended to confirm Kekulé's picture, although his explanation of the bonding structure has been improved by Pauling.

Then there are scientific discoveries of new experimental facts, such as a new substance or property of a substance. Such a discovery was that of Becquerel when he found that uranium fogged a photographic plate. Unlike a theory, which can be superseded by another theory, an experimental fact is permanent: if it is measurable (say, the density of a substance) it may be measured more accurately, but it cannot be replaced.

Often a new experimental fact leads to a new theoretical explanation: for example, Rutherford's explanation of how uranium fogs a photographic plate in terms of radioactive decay. Similarly, a new theory often leads to the discovery of a new experimental fact; for example, Einstein's General Theory of Relativity led to the discovery that light is influenced by gravity. In this way fact leads to theory, and theory to fact.

How a scientist comes to make a discovery may arise in a number of ways. It may be a chance observation that starts the scientist's mind thinking; something unexpected that happens in an experiment, such as when Fleming discovered the antibiotic properties of penicillin from the mould that spoilt his culture of bacteria.

Chance plays a part in many discoveries but, as Pasteur said, 'Chance favours the prepared mind'. A planned piece of research may follow on from a previous discovery, as Faraday's experiments to produce electricity from magnetism followed from Oersted's discovery that magnetism could be produced by electricity.

Or it may be something the scientist reads or hears that places something he, or she, knows in an entirely new perspective, such as the three bits of knowledge that clicked together in the mind of Lawrence Bragg and led to his new understanding of the X-ray patterns from a crystal. You can read about some of these discoveries here and in Chapter 4.

_ _ _ _ _ _ _ _ _ _ _ _ _ _ _ _

Which of the sciences do you think will produce the most exciting discoveries in the next ten years? What would you like to invent or discover?

_ _ _ _ _ _ _ _ _ _ _ _ _ _ _ _

It is only in the last seventy years that a way has been found to determine the positions of the atoms in a crystal. To do this, scientists use X-rays, which are waves of the same nature as light but with a wavelength that is ten thousand times shorter. Sir Lawrence Bragg, who wrote what follows, was the first person to realize that X-rays could be used to find out the structure of substances. His father, Sir William Bragg, invented the instrument with which investigations of crystal structure are carried out. Together they determined the structure of many substances and started the science of 'X-ray crystallography'.

SIR LAWRENCE BRAGG DESCRIBES THE START OF X-RAY ANALYSIS

'My interest in science started when I was at school, and I think the main reason was that my chemistry master taught in an interesting way. I went to school in Australia; I was born in Adelaide and my family lived there till we all came to England when I was eighteen years old.

'When we were in Adelaide, my father used to talk to me about his scientific ideas. He had gone to Adelaide as a young man to be Professor of Mathematics and Physics at the university, but he was so busy there building up the courses and practical work in the physics laboratory, and helping to develop Adelaide University, which was very new, that he never thought of doing research for

Figure 41
Sir William Bragg (1862–1942)

nearly twenty years. But when he was in his forties he was asked to give an address to the Australian Association for the Advancement of Science on the exciting new discoveries which were being made in radioactivity.

'Preparing his lecture he began to wonder whether the explanations of the way the rays from radium behaved were right, and he determined to check some of the properties for himself. So he got the University to buy him some radium, and started experiments on the α-rays from radium. They were brilliantly successful. Rutherford was tremendously interested because they fitted in so well with his theory that radioactive decay resulted from the breakdown of the atoms of radioactive elements. At this time Rutherford was trying to convince a doubting scientific world that, in radioactive processes, one element was changing into another, an idea quite in opposition to the existing doctrine that atoms were unchangeable.

'In two or three years my father became world famous as a pioneer in radioactivity. In 1908 we all came home to England because my father was made a Fellow of the Royal Society and was invited to become Professor of Physics at Leeds University. In England I went to Trinity College, Cambridge (figure 42). I started with mathematics but after a year my father thought it would be better if I switched over to physics. That was how I became a physicist.

'My father went on to study the other radiations coming from radioactive bodies. He was particularly interested in the γ-rays which, like X-rays, made a gas a conductor of electricity by turning the atoms or molecules into charged ions. Now my father showed by a series of very ingenious experiments that this is not a general effect all through the gas, but that certain gas atoms seem to get from the X-rays a kind of "direct hit" which sends an electron careering along at a vast speed, and it is these electrons which ionize the gas. One would never expect waves to act in this way, so my father came to the conclusion that γ-rays and X-rays were not waves but more like a lot of little bullets which hit an atom here and there: they were able to penetrate so far through solid bodies because they were electrically neutral, being a matched pair of positive and negative particles. I remember the very first time he told his great idea to me, just as we were boarding the old horse tram which ran down the main street near our home.

'He went on developing this "neutral-pair" hypothesis when he came to England, and had scientific fights with those people who believed in waves. So there was great excitement when, in 1912, a German scientist called von Laue published a paper with some beautiful photographs which, he claimed, showed that X-rays were undoubtedly waves (figure 43). He obtained these photographs by sending a narrow beam of X-rays through a crystal, and placing the photographic plate

Figure 42
Lawrence Bragg (1890–1971) at the time of his Nobel award in 1915.

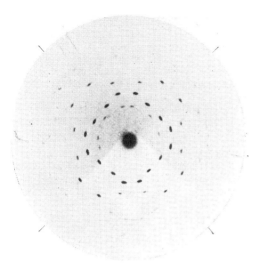

Figure 43
One of Von Laue's original photographs. Here the crystal is zincblende.
From Ewald, P.P. (1923) Kristalle und Röntgenstrahlen.

on the far side. When the plate was developed, it showed a number of spots in a pattern of the same symmetry as the crystal, and von Laue explained that the effect was caused by "diffraction" of waves by the regular lattice of the crystal.

'My father got von Laue's paper while we were on summer holiday on the Yorkshire coast, and we studied it together. I had just taken my degree at Cambridge and was then twenty-two. I, of course, was a tremendously warm believer in my father's theories, and we tried to find a way of explaining von Laue's photographs by something other than waves.

'However, when I returned to Cambridge for the autumn term and pondered over the photographs, I became convinced that von Laue was right in concluding that they were produced by waves. At the same time, I had an inspiration which led me to believe that when von Laue explained the peculiarities of his diffraction picture as being due to a complex set of wavelengths coming from the X-ray tube, he was on the wrong track, and that really the peculiarities were due to the way the atoms were arranged in the crystal. If this were so, then X-rays could be used to find out the arrangement of the atoms.

'It is worth while describing this inspiration in some detail because it shows how scientific ideas often arise. They come because one hears about a piece of knowledge from one source, and happens to have a quite separate piece of knowledge from another source, and somehow the two just click together and there is the new idea. In my case, it was a kind of treble-chance.

'First, J.J. Thomson had lectured to us about X-rays, and explained them as a wave-pulse in the ether caused by the electrons hitting the target in the X-ray tube and being stopped suddenly. Second, C.T.R. Wilson had given us very stimulating lectures on optics, including an analysis of white light which showed that one could think of it either as a series of quite irregular pulses or as a continuous range of wavelengths. Third, we had a little scientific society in Trinity, and at one of our meetings a member had read a paper about a theory that, in crystal structures, the atoms were packed together like spheres whose volumes were proportional to the combining power of the atoms. This theory had been proposed by Pope and Barlow and proved in the end to be quite wrong, but it suggested some very useful ideas. In science, a wrong theory can be very valuable, and much better than no theory at all.

'Hearing this paper, I realized that the atoms in crystals were arranged in parallel sheets. Anyone thinking about a crystal pattern would see this at once, but I had never thought about it before. So these three bits of knowledge were part of my background. When I was walking one day along the Backs at Cambridge – I can remember the place behind St. John's College – suddenly the three bits came together with a click in my mind. I suddenly realized that von Laue's spots were the reflections of the X-rays in the sheets of atoms in the crystal.'

(Bragg's explanation of the patterns of spots in von Laue's photographs is outlined in Chapter 9 of the *Handbook for pupils*. Bragg showed that there was a simple law connecting the wavelength of the X-rays, the spacing of the planes of atoms in the crystal, and the angles at which diffraction spots were obtained. He did some experiments to get the Laue patterns of potassium and sodium chloride, and he was able to use his law to work out the crystal structures of these compounds. A model of the structure of sodium chloride is also shown in Chapter 9 of the *Handbook for pupils*.)

'When I told my father about my results, he was of course very interested, and he at once started experiments to find out whether the rays which I had found to be reflected from crystal faces were in fact actual X-rays. When I wrote my paper I did not want to take this for granted, so I called the paper "The diffraction of short electromagnetic waves". I avoided mentioning X-rays, having been very much teased at Cambridge for upsetting my own father's theory!

'To make an accurate study of the waves, my father built what he called an "X-ray spectrometer" (figure 44). It was designed like the spectrometer one uses to examine the wavelengths of light. A fine beam of X-rays fell on a crystal face which could be set at a measured angle θ. The reflected ray was measured by what is called an ionization chamber, set at an angle 2θ. The X-rays made the gas in this chamber a conductor, and the charge which flowed through it was measured by the electrometer at the base. It was a beautiful instrument. My father was very good at designing scientific apparatus, and he had at Leeds a genius of an instrument-maker named Jenkinson. His experiments on γ-rays and X-rays had made him adept at accurate intensity measurements of these radiations. He soon satisfied himself that the reflected rays really were X-rays (figure 45).

'Although my father built the spectrometer in order to study the reflected waves, his instrument proved invaluable for use on crystal structure. For the earliest measurements with the spectrometer showed that it was a far more powerful way of finding out the pattern of the atomic particles in crystals than the roundabout and difficult method I had applied to the Laue photographs.

'I was at that time trying to interpret the Laue photographs produced by a diamond, and was quite bogged down. With my father's

Figure 44
The Bragg X-ray spectrometer. When in operation, it was essential to surround the X-ray tube with a lead-lined box for protection against the powerful X-rays.
Photographed by courtesy of The Royal Institution

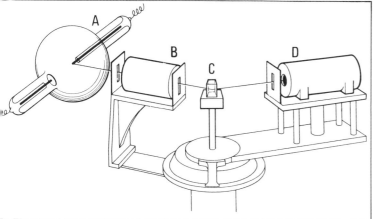

Figure 45
Diagram of the X-ray spectrometer. X-rays generated in the tube A pass through the slit system B on to the inclined face of a crystal C, and the reflection of the X-rays is measured in an ionization chamber D.

Figure 46
Linus Pauling, beside the α-helix, a configuration of the polypeptide chain of proteins.

spectrometer, it was possible to measure the reflections of the X-ray spectral lines from the diamond planes, and this led at once to a solution. The diamond structure aroused a great deal of interest and had a strong influence in convincing scientists of the value of "X-ray analysis", as it came to be called.

'I had a grand time in the holidays. My father's interests were still mainly in X-ray spectra, and he let me examine crystals with the X-ray spectrometer, and use measurements he had made to try to work out their structures. We worked furiously in 1913 and 1914, going back in the evenings to the deserted university to get more measurements. It was like discovering a goldfield with the nuggets just lying there to be picked up. One could not resist the temptation to pick up more and more, without a rest. I was very lucky. If it had not been my father who developed the X-ray spectrometer, I should never have been able to work with it'.

Do you find science exciting when you learn about new things? Can you imagine the enthusiasm of the research scientist making completely new discoveries? Have you ever been so interested in something that it has gripped your attention day and night for weeks?

The next three sections are based on personal interviews with famous scientists who have used X-ray analysis to investigate the structure of naturally occurring large molecules.

LINUS PAULING – BONDING AND STRUCTURE

Linus Pauling was born in Portland, Oregon, U.S.A., in 1901. He received his doctorate from the California Institute of Technology, where he worked for over forty years. He is an active campaigner against the use of nuclear weapons and against war. He was President of the Linus Pauling Institute for Science and Medicine, San Diego, California, 1973–1975. Apart from Madame Curie, he is the only person ever to have won two Nobel prizes: the Chemistry Prize in 1954 and the Peace Prize in 1962.

He was awarded the Chemistry prize 'for his research into the nature of the chemical bond and its application to the elucidation of the structure of complex substances' (figure 46).

The way of discovery

When Linus Pauling was eighteen, he worked as a paving-plant inspector in Oregon during an interlude in his student career. His job was merely to test the temperature and composition of the mixes about once an hour. To pass the time he tried working out a theory that related the structure of substances to their magnetic properties. Chemical structure, the arrangement of atoms in substances and the ways in which they are linked together, has remained his predominant interest ever since. But his contributions to scientific knowledge have been many and varied — from proposing the electronegativity scale in 1931 to his chemical explanation of anaesthesia in 1961.

When Pauling began research at the California Institute of Technology in 1922, he made X-ray diffraction studies of many substances, measuring such things as the bond angles, interatomic distances and ionic radii. 'I enjoy experimental work but I tend to be rather impatient. I prefer simple experiments and try to find an experimental technique that, once the apparatus is built, can be applied to a large number of substances.'

At this time X-ray techniques were very much less advanced than they are now. Pauling determined the geometrical structure of many complex substances using a method which, in his usual way, he worked out very largely for himself. He calls the method 'stochastic' — guessing the truth.

'Here, for example, I have a crystal of pseudobrookite (Fe_2TiO_5). I can obtain sure information by X-ray diffraction — size of the unit, space symmetry — so that there are some simple restrictions on the possible structure. Keeping within these restrictions, what would be the most reasonable way for two atoms of iron, one of titanium, and five atoms of oxygen, to link together? Often, after working with pencil and paper and simple ball and stick models, I end up with only one possible structure. I make some calculations based on this structure and test them experimentally. If the calculations and experimental results agree, then I accept the structure. If not, that's too bad; I usually drop it.'

Using this method, which requires extraordinary understanding of how atoms behave, Pauling determined the structure of topaz, mica, chlorite, and many other minerals, with all the atoms located in their correct places. One of his failures was sodium dicadmide ($NaCd_2$). One of his students eventually used the stochastic method to work this out. It proved to be the most complex inorganic structure known, with 1192 atoms in the unit cell.

In 1937, Pauling began to explore the geometrical structure of proteins. He was unsuccessful at first, because he had too few data to go on. During the next ten years, his research team made X-ray diffraction studies of amino acids and peptides, of which proteins are composed; they measured bond distances and bond angles. In 1948, using these new data, Pauling conceived the idea that the fibrous protein molecule of α-keratin was folded into a helical structure. This idea of a helix was to prove fundamental in understanding the structure of a number of complex biological substances.

'For solving problems that initially defeat me I deliberately make use of my unconscious mind. I think about the problem for about half an hour in bed and then go to sleep still thinking about it. I do this, perhaps, for several nights, and then forget about it all together. Months or sometimes years later, as with the structure of α-keratin, the answer pops into my head.' He chooses, of course, problems to work on that he thinks he will ultimately be able to solve. This is a matter of judgment — the ability to sort through a mass of ideas quickly and fix on the interesting ones. 'One of the most important things is to be willing to accept new ideas and to look at problems from a fresh perspective.'

DOROTHY HODGKIN — X-RAY ANALYSIS OF ORGANIC MOLECULES

Dorothy Hodgkin was born in Egypt on 12 May 1910. After graduating in chemistry from Oxford in 1932, she spent two years doing research at Cambridge, and then returned to Oxford where she has been ever since. She married in 1937 and, along with her research studies, has brought up a family of three. 'Looking after children,' she says, 'is the main difficulty that faces a woman who wants to do research.' She is only the third woman ever to win a Nobel Prize in Chemistry — Madame Curie and her daughter Irene Joliot-Curie were the others. She was awarded the Chemistry Prize in 1964 'for her determination by X-ray techniques of the structure of important biochemical substances' (figure 47).

'Crystals are the most obvious thing to like when you're starting chemistry and I've kept with them ever since.' Dorothy Hodgkin realized, during the lectures she attended at Oxford at the beginning of the thirties, that evidence for the chemical formulae of such organic compounds as sterols and strychnine was very sketchy and that the geometrical arrangement of the atoms was virtually unknown. Here, it seemed, were problems that X-ray analysis might be able to solve. She had been interested in the technique of X-ray analysis even before going to Oxford. When it came to research, it was the X-ray study of organic substances that attracted her.

Figure 48
X-ray photograph of a crystal of Vitamin B_{12}.

Figure 47
Dorothy Hodgkin.
Keystone Press Agency Ltd.

She took an X-ray photograph of insulin as long ago as 1935, the second protein crystal ever to be X-ray photographed. She remembers this as the most exciting moment in her career: 'I developed the photographs late at night and walked elated round the streets of Oxford before going to bed. Then I woke up early, worried that the crystals might not be insulin after all. I slipped round to the laboratory before breakfast to test that I really had protein crystals.' At the time, too little was known about proteins for her to be able to interpret the photograph.

Of the many complex structures that Dorothy Hodgkin has examined, two in particular have established her reputation: that of penicillin, which she determined during the Second World War; and that of Vitamin B_{12} which she determined during the early fifties.

Her general approach was to discover the geometrical arrangement of the atoms in space, and thus, the precise way in which the atoms were linked together which established the actual chemical structures of the compounds. In the case of penicillin, parts of the structure were found by chemical methods before the compound was crystallized; it was, however, the X-ray measurements which showed clearly how these parts were joined together. In the case of Vitamin B_{12} even the number and the precise chemical nature of the atoms present were found through the X-ray crystal structure determination.

The way of discovery

Dorothy Hodgkin's success, combined with that of many others, did much to establish X-ray crystallography as a method of determining the chemical structure of complex molecules as well as reaffirming its use for establishing molecular geometry when the chemical structure is known. Nowadays analyses of this kind have been made much easier by greater use of computers. Mrs Hodgkin dismisses much of her achievement as 'pig-headedness' which led her to pursue difficult structure analysis before the techniques had been developed that were to make them easy.

'The kind of skill needed in X-ray analysis is a feeling for the way in which atoms are arranged in a molecule, being able to look at a series of X-ray photographs to see how the atoms might fit together. It needs imagination, too, to visualize the various possible arrangements, and it needs judgment born of experience to select those possibilities which might lead to an answer. Sometimes the answer comes out very quickly: but with complex molecules such as Vitamin B_{12}, it may take years — working out the arrangement of atoms piece by piece and then seeing how the pieces fit together in the whole molecule.

'In each analysis there may be several interesting stages — getting an awkward substance to crystallize; seeing just what one is up against from the first X-ray photographs; and a point, often long before the end, when one knows that the answer is in sight. Of course, X-ray work has its limitations: one cannot necessarily predict from the arrangement of the atoms how a substance will behave in a chemical reaction. But it is a useful tool and can give one the position of the atoms in space more quickly and with greater certainty than any other method.'

JAMES D. WATSON — THE STRUCTURE OF DNA

James D. Watson was born in Chicago, Illinois, U.S.A., in 1928. He studied biology at the Universities of Chicago and Indiana and, after a year's research in Copenhagen, came to the Cavendish Laboratory in Cambridge in 1951. He is at present Professor of Biology at Harvard University. He is also Director of the Coldspring Harbor Laboratory, where he has been since 1968.

In 1962, he shared the Nobel Prize in Medicine and Biology with F.H. Crick and M.F.H. Wilkins — 'for their discoveries concerning the structure of nucleic acids and its significance for information transfer in living material' (figure 49).

In the nucleus of each living cell are millions of long thread-like molecules of a substance called deoxyribonucleic acid (DNA). For many years it had been thought that these molecules were in some way responsible for transferring hereditary information from parent organisms to the next generation. It was not understood how. In the early fifties, Todd and his co-workers at Cambridge worked out the chemical composition of DNA, but nothing was known about the

Figure 49
Nobel prize winners in 1962; from left to right, Maurice Wilkins, Max Perutz, Francis Crick, John Steinbeck, James D. Watson, and John Kendrew.
Keystone Press Agency Ltd

geometrical arrangement of the atoms. And it is often the geometrical structure of biological substances that provides the clue to the way in which they function.

When this young American biologist, James D. Watson, came to the Department of Molecular Biology in the Cavendish Laboratory at Cambridge, he was interested in genetics and wanted to discover the structure of DNA. At the time, most people thought the problem was too difficult. Sharing a room with Watson at the Cavendish was Francis Crick who was working on a thesis. Crick wasn't particularly interested in the subject of his thesis. He was fascinated by the structural problem confronting Watson.

The backbone of the DNA molecular chain consists of an alternating sequence of phosphate groups and sugar groups. A side group of atoms called a base projects from each sugar group. There are four different types of base. 'For hours every day Crick and I talked about how these groups and bases could fit together in a sensible geometrical shape. We built simple ball-and-stick models, using the same kind of conjectural approach devised by Pauling, but for a long time we seemed to be getting nowhere'. Pauling, who had previously shown that many proteins have a helical structure, tried briefly to work out the structure of DNA himself, but he had no success. Another person studying the same problem was Maurice Wilkins at London University. He was trying to do it experimentally from X-ray diffraction pictures; these pictures suggested that the DNA molecule had a symmetrical structure.

After two years of discussion, of trial-and-error rearrangement of models, of grappling with a base whose structure had been wrongly represented in the textbooks, it occurred to Watson and Crick that the framework of the DNA structure might be that of a double helix. 'After that we had the structure of the whole molecule virtually worked out in less than a week.'

It consisted of two chains of alternating phosphate and sugar groups spiralling round each other to form a double helix with the two bases filling in the centre (figure 50). Although much has still to be understood, the discovery of the structure of DNA was a great step towards explaining the biology of cells in fundamental molecular terms. 'It is impossible', says Watson, 'to distinguish between the separate contributions of Crick and myself to solving this problem; but if either of us hadn't been there, then it wouldn't have happened'.

You can read something of the story of the discovery of the structure of DNA in James Watson's book *The double helix*, published in 1968 by Weidenfeld and Nicolson. It is an exciting personal account of the events leading up to the discovery, and of the people involved.

Figure 50
Molecular model of DNA (deoxyribonucleic acid).
By courtesy of Professor M.H.F. Wilkins, Department of Biophysics, King's College, London

The way of discovery

Chapter 4
DAVY AND FARADAY

HUMPHRY DAVY 1778–1829

A lively mind

Humphry Davy was born at a time when brilliant and dedicated men throughout Europe were laying the foundations of modern science. Davy was outstanding among them. Before he was thirty, he had discovered six new elements and, in so doing, had established electricity as one of the chemist's most useful experimental tools. He made several other important discoveries and he invented the miner's safety lamp. One of his greatest talents was his ability to popularize scientific discovery (figure 51).

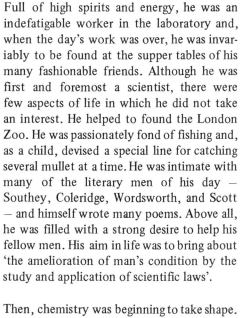

Figure 51
Humphry Davy (1778–1829) as a young man, engraved by Roffe from a painting by Howard.
Science Museum, London.

Full of high spirits and energy, he was an indefatigable worker in the laboratory and, when the day's work was over, he was invariably to be found at the supper tables of his many fashionable friends. Although he was first and foremost a scientist, there were few aspects of life in which he did not take an interest. He helped to found the London Zoo. He was passionately fond of fishing and, as a child, devised a special line for catching several mullet at a time. He was intimate with many of the literary men of his day – Southey, Coleridge, Wordsworth, and Scott – and himself wrote many poems. Above all, he was filled with a strong desire to help his fellow men. His aim in life was to bring about 'the amelioration of man's condition by the study and application of scientific laws'.

Then, chemistry was beginning to take shape. From the end of the seventeenth century, it changed gradually from a mixture of philosophy and recipe-mongering into something more truly scientific, as chemists began to develop theories based on experiments that had logical connections one with another.

Boyhood and youth

Humphry Davy was born in Penzance in 1778. He was educated at the local grammar school but he did not work hard and was often in trouble. But he lived in a prosperous region where tin and copper mines were a centre of technical progress. His interest in what substances were made of, and how they could be put to practical use, was early awakened.

Soon after he left school, at the age of fifteen, his father died and he had to support his mother and her four other children. He was apprenticed to Mr Bingham Borlase, a surgeon and apothecary. At the age of nineteen he began his experimental study of chemistry, working in his spare time in an attic.

During his apprenticeship, Davy's gifts were brought to the attention of Dr Thomas Beddoes of Bristol. Dr Beddoes was impressed and offered Davy a post as superintendent of a laboratory he was setting up to study the medicinal effects of gases.

At Bristol Davy set to work to examine the properties of nitrous oxide (now called dinitrogen oxide), a gas which had been discovered some years earlier by Priestley. Davy had already made small amounts of nitrous oxide at Penzance, but at Bristol he made it in its pure form and examined its chemical composition, obtaining an accurate estimate of the proportions of nitrogen and oxygen.

There had been some argument about the medical properties of nitrous oxide, and one theory held that it possessed the power of spreading disease. Davy decided that the only way to test this was to try it on himself, and breathed two quarts of it from a silk bag. He found it was safe enough, but was astonished at the effect it produced. It was like being drunk or semi-delirious, but in a very pleasant way. Many other people tried it: some became so uncontrollably hilarious that the gas got the name of 'laughing gas'.

The most striking effect of the gas that Davy noticed was that it suppressed physical pain. (It stopped one of his wisdom teeth aching while he breathed it.) Davy wrote 'As nitrous oxide . . . appears capable of destroying physical pain, it may probably be used with advantage during surgical operations.'

Unfortunately Davy's suggestion was not adopted until almost fifty years later when an American dentist Horace Wells used it as an anaesthetic: but, by then, ether had come into use for this purpose and it was some time before 'laughing gas' became established as an anaesthetic for minor operations. Even so, Davy had made an interesting discovery and gained a reputation as an experimenter. He pursued his studies of gases and was lucky not to kill himself with some experiments on carbon monoxide.

Davy nearly killed himself when studying gas. What safety precautions would you take when making and testing a gas which you had not met before?

Triumph in London

Davy's next move was to the Royal Institution in London, and it was there that he established himself as one of the foremost scientists of his day. The Royal Institution had been founded in 1799 by Count Rumford a man distinguished in public affairs. Rumford's purpose in starting the Institution was to apply science to such everyday needs as the preparation of 'cheap and nutritious foods for feeding the poor' and 'improving the construction of cottages, and cottage fireplaces and kitchen utensils'.

The Institution also aimed at fostering an interest in science among the well-to-do classes and among artisans, for people were becoming curious about what science could do. Davy was recommended to Rumford as a promising young man and, with the consent of Dr Beddoes, Rumford's friend, was appointed lecture assistant at the Institution. The first professor of chemistry, Dr Garnett, had begun well, but after serious illness and several disputes with Rumford, had to resign. Davy was appointed by Rumford to take Garnett's place.

From his very first lectures in 1801 Davy was a great success. Hundreds of people packed into the lecture room to hear this eloquent young man talking about chemistry and its applications (plate 5). He was always deeply excited by his subject and managed to communicate this excitement to his audience. Wealthy and influential people came to listen to him, and his charm and good looks soon made him sought after by the fashionable leaders of society. When asked how the clever men in London compared with Davy, the poet and philosopher Coleridge replied: 'Clever men? Our own Humphry could eat them all.' Such praise greatly pleased the ambitious young Davy but, if it made him rather vain, it did not distract him from his scientific work.

Davy's laboratory was in an underground room below the Royal Institution; and though he was neat and methodical in the lecture room, in his laboratory he was quite another person. He worked very fast; invariably the laboratory was in a state of chaos, and he often carried on several quite unconnected experiments at the same time. In contrast to the painstaking Michael Faraday, who was later to become his assistant, he was all brilliance and dash. Yet, despite his unconcern for detail, he achieved some remarkable results. For some time his experimental research, selected by the Managers of the Institution, was on such subjects as the tanning of leather or the analysis of minerals. But by 1806 he was able to take up once more some interrupted work on electricity.

Electricity

During the eighteenth century there had been much progress in the study of electricity but only static electricity was known. This was produced by friction, often continuously by a machine.

In 1791 Galvani's experiments on frogs and animal electricity attracted attention. Then, in 1800, details of Volta's electric battery or pile were published. A set of alternate plates of zinc and silver separated by damp linen or paper produced electricity spontaneously without external action. Almost immediately two English scientists Nicholson and Carlisle made a very remarkable observation when they connected a wire from the bottom plate of a Voltaic pile to a drop of water on the top plate: they saw that bubbles of gas were formed (figure 52).

Like many other scientists, Davy followed up this experiment. He had already found in 1800, while still at Bristol, that the amounts of hydrogen and oxygen produced from water by electrolysis were in the proportion 2:1 by volume, which is the same as the proportion in which hydrogen and oxygen combine to form water. When the composition of even as simple a substance as water was still in dispute, this discovery was significant. With caustic potash (potassium hydroxide), electrolysis produced hydrogen and oxygen – exactly the gases obtained from water. This result was a surprise and for a long time there was no explanation for it.

Because electricity produced chemical reactions, Davy reasoned that electricity might very well be the cause of chemical affinity, and that in some form or another it might be an essential constituent of all matter. This was a new and fundamental idea, the full meaning of which has only been worked out in the twentieth century, but it had a profound effect on science right from the time of its discovery (figure 53).

A year later, in 1807, Davy crowned his electrical work with a discovery which really deserves to be called sensational. He came back to the electrolysis of potash, using, not a solution, but some solid potash, barely moist. At the point of contact of the platinum wire there appeared small globules of a shining white metal which rapidly burnt in air.

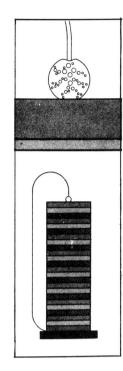

Figure 52
Diagram of a Voltaic pile built by Nicholson and Carlisle consisting of thirty-six silver half-crowns alternating with thirty-six zinc discs, separated by paper soaked in salt water. With platinum wire they connected the bottom plate of the pile to a drop of water on the top plate. Immediately gas bubbles began to form which proved to be oxygen and hydrogen. Davy said: 'The origin of all that has been done in electrochemical science was the discovery by Nicholson and Carlisle of the decomposition of water on 30 April 1800.'

Figure 53
Some of the original apparatus with which Davy carried out his electrochemical experiments. The battery (at the rear of the photograph) is a modified version of Volta's original pile.
Reproduced by permission of The Royal Institution

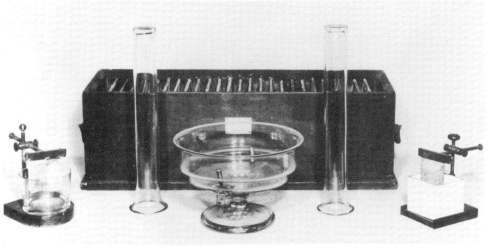

This was potassium. In his notebook he wrote: 'October 19 (1807) When potash was introduced into a tube having a platina wire attached to it so (this refers to the rough drawing in the first paragraph of figure 54), and fused into the tube so as to be a conductor *i.e.* so as to contain just water enough though solid — and inserted over mercury. When the Platina was made negative, no gas was formed and the mercury became oxydated and a small quantity of the alkaligen (potassium) was produced round the platinum wire as was evident from its giving inflammation by the action of water. When the mercury was made the negative, gas was developed in great quantities from the positive wire and none from the negative mercury and this gas proved to be pure oxygene. Capital Experiment proving the decomposition of potash.' (Figure 54.)

Within three days, he had isolated another metal from caustic soda (sodium hydroxide) by the same method. Eventually he called these new metals potassium and sodium. But the pace at which he had been working proved too much for him, and he collapsed from the strain.

After a serious illness lasting several months, Davy returned to his laboratory to continue work with the Voltaic pile which, by his successful decomposition of potash and caustic soda, he had shown to be such a useful tool for the chemist.

Following up suggestions of the Swedish chemist Berzelius, he used a mercury electrode to obtain amalgams of two metals from the minerals baryta and strontia. By distilling the mercury, he obtained the free metals which he called barium and strontium. He could not obtain pure calcium or magnesium by this method, but the connections were clear enough for his assertion of four new elements to be justified. Six new elements within two years was good going!

Oxymuriatic acid

Davy's next important research was of a different kind. His work with electricity had produced substances which nobody had seen before. Davy now showed that a substance every chemist knew very well as a chemical compound was, in fact, an element. This was chlorine.

Chlorine had been discovered by Scheele in 1774. Its compound with hydrogen, the substance we call hydrochloric acid, had been well known for hundreds of years. It was called *muriatic* acid (from the Latin word for 'brine' — *muria*).

Lavoisier had believed oxygen to be present in all acids; and in fact the name Lavoisier gave this gas means 'acid-maker'. What, then, was the nature of chlorine? It was made by oxidizing 'muriatic' acid. Thus, when Lavoisier called it 'oxymuriatic' acid, everybody agreed.

In 1810 Davy decided that there was only one argument in favour of believing 'oxymuriatic' acid to be a compound containing oxygen. This was that all other acids that were known were made from oxides, for example, from oxides of nitrogen or sulphur. He carried out experiments aimed at extracting the oxygen from compounds containing it. He heated the 'oxymuriatic acid' with white-hot charcoal; he passed electric sparks through it; he examined its compounds with tin and phosphorus and could obtain no oxygen compounds from them. He pointed out that, although oxygen could be obtained from oxymuriatic acid, this only happened when water was present. After considering such evidence as this, Davy modestly concluded: 'There may be oxygen in oxymuriatic acid, but I can find none.'

Figure 54
Davy's notes of the famous experiment in which he decomposed potash and was led to the discovery of the metal potassium.
Reproduced by permission of The Royal Institution

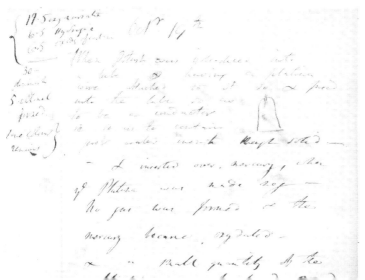

Thus a prominent defect in Lavoisier's ideas about acids was removed, for it was recognized that muriatic acid (that is, hydrochloric acid) was a compound of chlorine and hydrogen containing no oxygen. This led the way to a new idea that an acid is a compound containing hydrogen and that the hydrogen can be replaced by a metal to form a salt.

Man of fame

In 1812 Davy married a rich widow, a cousin of Sir Walter Scott. He was thirty-three years old and at the height of his fame. A knighthood was conferred on him by the Prince Regent. For several years he had been a prominent member of the Royal Society which was (and is) the leading scientific society in Britain. Later Davy was to become its president. Shortly after marrying, he resigned his professorship at the Royal Institution, but retained an honorary post and continued to do research in the laboratories.

A new assistant

One day in 1813 an extraordinary thing happened at the Royal Institution. The laboratory assistant William Payne was found fighting with the instrument maker, Mr Newman. Payne was dismissed and someone had to be found to take his place. Humphry Davy recommended a young man who wished to be considered for work at the Institution. Davy reported: 'His name is Michael Faraday. His habits seem good, his disposition active and cheerful, and his manner intelligent.' Faraday was appointed at a salary of twenty-five shillings a week, with rooms at the Institution. He was then twenty-one. How did Davy know about him?

Michael Faraday was the second son of a Yorkshire blacksmith who had moved to London. Faraday senior and all his family were members of a strict religious sect called Sandemanians which had separated from the Presbyterian Church of Scotland in 1730. The small London group of Sandemanians devoted every Sunday to religious service at a meeting-house in one of the worst slums of London. The elders of the group held the services and took turns to give the sermons.

Faraday's family was poor. His father had difficulty in finding regular work, particularly when his health began to fail. In 1801 the Faradays were forced to ask for public relief and young Faraday's share worked out at one loaf of bread, which had to last him a week.

When he was thirteen Faraday became errand boy to a bookseller called Riebau, for whom he delivered morning newspapers and collected and delivered books. Riebau soon came to like the boy and, knowing that the Faraday family could not possibly afford to pay a premium, offered to accept him as an apprentice bookbinder without fee.

Towards the end of his apprenticeship, Faraday went to scientific lectures by a Mr Tatum of Fleet Street. Tatum's lecture courses eventually resulted in the formation of Birkbeck College, London.

Early in 1812, a customer of Riebau called Dance, who was a member of the Royal Institution, gave him tickets to hear four lectures by Humphry Davy. Faraday took careful notes of all he saw and heard, illustrated the notes and bound them. These notes were to serve him well.

In October 1812 Faraday's apprenticeship with Riebau came to an end, and reluctantly he left to work as a journeyman bookbinder for De La Roche, a rather hasty-tempered French immigrant. Although he could now make some contribution to the family finances, Faraday was no longer as happy as he had been as an apprentice. His desire to find some employment of a scientific nature made him more and more restive.

He had previously written to Sir Joseph Banks, President of the Royal Society, asking whether the Society could find him laboratory work, however menial, but the letter was returned marked 'no reply'. Encouraged by Dance and the Riebaus, Faraday next wrote to Humphry Davy, asking for a job as a laboratory assistant. He sent with the letter his bound notes of the four lectures by Davy. On Christmas Eve, 1812, Davy replied saying how pleased he was with the notes and arranged an interview, but pointed out that Faraday would be much better off as a bookbinder, adding that at the moment there was no vacancy at the Royal Institution.

Two days later, much to his surprise, Faraday received another letter from Davy asking him to write a fair copy of some of his notes for the next issue of the *Quarterly Journal*. Davy could not write them himself because he had injured his eye in an explosion resulting from an experiment with nitrogen chloride. (This dangerous substance had been discovered in 1811 by Dulong, who continued to investigate it despite the loss of one eye and three fingers in its first preparation.)

Faraday, quickly learning to read Davy's untidy and almost illegible writing, rewrote the notes in his own immaculate hand. For the three days' work he received thirty shillings and, at the same time, consolidated the good impression he had originally made. Shortly afterwards the fight in the laboratory took place, William Payne was dismissed, and Davy sent for Faraday. He offered him Payne's former position and asked him also to act as his personal laboratory assistant and secretary. Faraday started his work at the Royal Institution on 6 March 1813.

Davy had planned a scientific tour of Europe starting in the autumn and he asked Faraday to accompany him as his assistant and secretary. During the tour Davy showed that a newly isolated black solid was an element similar to chlorine and he proposed that it should be called iodine. The tour lasted eighteen months. On returning to England, Davy turned his attention to the problem of explosions in coal mines, while Faraday was re-engaged at the Royal Institution.

How many of the elements which were first discovered or identified by Davy could not have been obtained without using electrolysis?

The safety lamp

The miner's safety lamp is one of those blessings of science which seems to have had no unpleasant consequences. Some inventions are made by luck or a quick idea. Davy's invention in 1815 was made by careful study of a technical problem.

Coal mining had been for years one of the country's most important industries. The miner found his way by open lights, candles, or oil lamps. As mines went deeper, explosions caused by gas became more dangerous. A society was formed to try to improve conditions and Davy was asked to help. He replied as follows: 'It will give me great satisfaction if my chemical knowledge be of any use in an enquiry so interesting to humanity, and I beg you will assure the Committee of my readiness to cooperate with them in any experiments or investigations on the subject.'

Davy started by visiting mines. Then he began to study flames and explosions in his laboratory. He found that gas explosions would not pass through narrow tubes, especially if they were made of metal. He reasoned that this must be because metal (which is a good conductor of heat) carried the heat of the burning gases away from the part already on fire. The next part, therefore, would not get hot enough to burn.

He then tried a wire gauze, which can be thought of as a lot of short narrow metal tubes side by side. This he found to be very effective. The inflammable gas can be burning on one side of the gauze but the gauze conducts the heat away so well that the gas on the other side is not raised to a temperature at which it will catch fire. Davy constructed lamps in which the flame was surrounded by a metal gauze (plate 6).

It was found that the Davy lamp not only burned safely but also gave a warning. This was because the inflammable gas which passed through the gauze to the inside of the lamp burned with a bright flame which looked quite different from the ordinary oil flame.

The lamp saved countless lives. It meant safer work, and therefore more productive work. The coal-owners were saved so much money that they gave Davy a present of a magnificent set of silver plate; but Davy, who had refused to accept any money payment for his work, left even this, in his will, to the Royal Society. He wanted it to pay for an annual medal awarded for any important chemical discovery.

(George Stephenson, later to be famous for his work as a railway engineer, also came close to designing a safety lamp along the same lines but his work was incomplete and unscientific. The real credit goes to Davy.)

Davy's last years

Davy's last major piece of work was carried out at the request of the Admiralty. He was asked to find a way to prevent the corrosion of the copper used to sheath ships' hulls. Davy suggested that bars of a more reactive metal such as zinc should be attached to the hull. This worked, but Davy had not realized that the products of the corrosion of the copper were poisonous to barnacles and seaweed. As the copper stayed bright and no longer poisoned the barnacles, they multiplied considerably and slowed down the ship.

Davy's last years were not very productive. He travelled restlessly and died in Geneva in 1829.

Which of Davy's discoveries and inventions do you think was the most important?
Why do you think Davy was such a successful scientist?

MICHAEL FARADAY 1791–1867

The first electric motor

In 1821, when he was thirty, Faraday made his first important discovery (figure 55). It was in the subject of electromagnetism, a science in which he was soon to lead the world. The year before, Oersted of Copenhagen had shown that if a wire carrying an electric current was held above, and parallel to, a compass needle, the needle was deflected.

Oersted's discovery was followed up by many scientists, in particular by Faraday and Wollaston in London, by Henry in America, and by Ampere in Paris. Oersted had said that the magnetic forces seemed to act in circles round the wire. It was Ampere who showed that parallel currents flowing in the same direction attract one another and so worked out the laws relating the current to the direction in which a magnetic needle moves.

The first efforts of the British scientists were directed to using a magnet to rotate a wire carrying a current. Wollaston was convinced this could be done, despite the doubts of Davy with whom he discussed it, but he was unable to devise a workable experiment.

Faraday, who knew of Wollaston's idea, eventually set up an ingenious apparatus in which one end of a bar magnet protruded through the surface of some mercury in a cup. A straight piece of copper wire was suspended

Figure 55
Michael Faraday in 1852, drawn by George Richmond.
Reproduced by permission of The Royal Institution

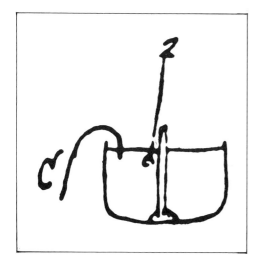

above the magnet with one end (to which was attached a small piece of cork to float it clear of the magnet) dipping into the mercury.

A second piece of wire was hooked over the rim of the cup and also dipped into the mercury. If the two wires (marked Z and C in the sketch, figure 56) were attached to the poles of a battery: an electric circuit was formed. The wire Z then began to rotate round the magnet. If the magnet was turned the other way up or if the poles of the battery were transposed, the wire rotated in the opposite direction.

Faraday promptly published his experiments in the *Quarterly Journal*. This publication, without any acknowledgement of the help he had received from the ideas of Davy and Wollaston, brought Faraday some criticism and made Davy rather cautious about discussing his ideas with him. Eventually Faraday apologized to both Davy and Wollaston, and after his first excursion into electricity he returned to chemistry (figure 57).

Chemistry

For nearly six years Faraday investigated the properties of steel. Considering the primitive equipment at his disposal, he produced some excellent samples, some of which were made into razors for his friends.

During the same period Faraday also investigated chlorine and its compounds in which

Figure 56
Faraday's original sketch of the first electric motor. *Reproduced by permission of The Royal Institution*

Figure 57
Faraday's Magnetic Laboratory, from a watercolour by Harriet Moore. The laboratory has been reconstructed and may be visited. *Reproduced by permission of The Royal Institution*

he had always been particularly interested. In 1823 he prepared chlorine hydrate — white crystals formed by cooling a solution of chlorine in water. At Davy's suggestion, he heated the crystals in one end of a sealed tube and noticed that oily drops collected in the cold part of the tube. Faraday recognized the drops as liquid chlorine and noted how rapidly they evaporated when the tube was broken open.

By similar methods he subsequently liquefied sulphur dioxide, carbon dioxide, and ammonia. The tubes occasionally burst, and he suffered a serious eye injury. Again his publication of the results was thought to have made too little recognition of Davy's ideas and relations between the Director and his assistant became more strained.

Later in 1823 Faraday was elected as a Fellow of the Royal Society and in 1825 he was appointed Director of the Laboratory at the Royal Institution.

Faraday soon put new life into the Institution. He helped to introduce evening meetings and demonstrations. These later became the famous Friday Evening Discourses which continue today. In 1826 he helped start the equally famous Christmas Lectures for young people (plate 7).

The responsibilities of directorship did nothing to hinder Faraday in his work. On the contrary, he intensified his research, his restless experimental curiosity grappling with one scientific problem after another. Nothing was too small to interest him, nothing too great to deter him. Nature was continually throwing up problems to be solved, and the answers, he believed, were always to be found through careful experiment. During his first years in office he investigated the liquid residue in cylinders of compressed gas which were used to light some of the richer homes in London. The gas was made by heating whale oil and was supplied in iron containers at a pressure of about thirty atmospheres.

The liquid proved to be a mixture of substances, one of which Faraday separated, purified, and analysed. This substance was previously unknown and he called it 'bicarburet of hydrogen'. We now call this substance *benzene*. It was first obtained in quantity from coal-tar, and is an important substance in the dye industry.

This first year of office also found Faraday starting a long series of experiments, sponsored by the Royal Society, to find better glass for optical work. His findings were not specially remarkable, but one glass, consisting chiefly of lead borosilicate, he used in an important discovery twenty years later — that a beam of polarized light passed through the glass was affected by an electromagnet.

———————————————

Do you think that it is important for scientists to explain their discoveries to the public? Do scientists today spend enough time interesting the public in science by writing for popular magazines, appearing on television, or speaking on the radio?

———————————————

Electromagnetism

Faraday is perhaps most famous for his discoveries in electromagnetism. In the summer of 1831 the Royal Society agreed that the experiments on glass should be suspended and Faraday began to study electromagnetism in earnest. He became convinced that since electricity could produce magnetic effects magnetism could be used to produce electricity.

In 1831 Faraday wound two separate coils of insulated wire round opposite sides of an iron ring and carried out the experiment described in the following excerpt from his notes (figure 58).

'August 29th, 1831. Experiments on the production of Electricity from Magnetism. Have had an iron ring made (soft ring) iron round and 7/8 inches thick and ring 6 inches in external diameter. Wound many coils of copper wire round one half, the coils being separated by twine and calico — there were 3 lengths of wire each about 24 feet long and they could be connected as one length or used as separate lengths. By trial with a trough each was insulated from the other. Will call this side of the ring A. On the other side but separated by an interval was wound wire in two pieces together amounting to about 60 feet in length, the direction being as with the former coils. This side call B. Charged a battery of 10 pairs of plates 4 inches square. Made the coil on B side one coil and connected its extremities by a copper wire passing

Figure 58
The original iron ring used by Faraday in his experiment to produce electricity by magnetism.
Reproduced by permission of The Royal Institution

Plate 1 *(top)*
Dalton collecting marsh gas (methane) by stirring the bottom of a pond. It was through his study of gases that Dalton was led to his atomic theory. The painting is by Ford Madox Brown.
Manchester City Council

Plate 2 *(centre left)*
'Radium', a contempory cartoon of Marie and Pierre Curie, 1904.
The Cribb Collection, by courtesy of the Royal Institute of Chemistry

Plate 3 *(bottom left)*
An aerial view of the Dounreay 250 MW prototype fast reactor which commenced operations in 1974. It is expected that fast reactors will be major power producers in the 1990s.
U.K.A.E.A.

Plate 4 *(bottom right)*
Nuclear Enterprises Ltd, Atomat nucleonic gauges for measuring the thickness of tyre cord at an Avon Rubber Company factory.
U.K.A.E.A.

Plate 5 *(above)* A cartoon by Gillray, 22 May 1802, of a lecture at the Royal Institution in 1801: 'Scientific Researches! — New Discoveries in PNEUMATICKS! — or — an Experimental Lecture on the Powers of Air'. The subject is the gases of the air. Davy, the young man behind the bench squeezing a pair of bellows, had already contributed to knowledge of gases through his work on nitrous oxide. Professor Garnett, Professor of Natural Philosophy and Chemistry at the Royal Institution, is administering gas to a member of the audience. Count Rumford, founder of the Royal Institution, is standing on the right. *The Cribb Collection, by courtesy of the Royal Institute of Chemistry.*

Plate 6 *(above)* Two versions of Davy's safety lamp. *Photographed by courtesy of the Royal Institution*

Plate 7 *(below)* A painting by Alexander Blakely of Michael Faraday lecturing to the Royal Institution, in the presence of the Prince Consort and the young princes, 27 December 1855. The Royal party can be seen on the left. *Photographed by courtesy of the Chemical Society.*

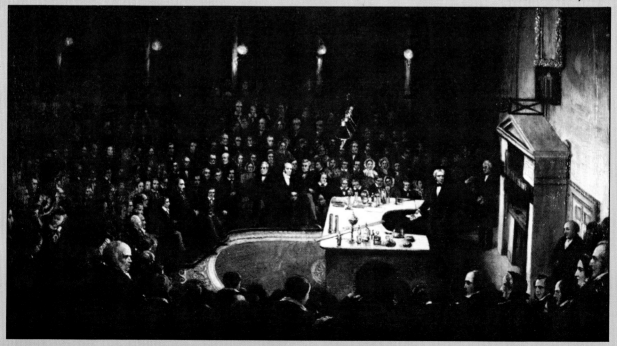

Plate 8 *(left)*
An electric arc furnace.
British Steel Corporation

Plate 9 *(right)*
Coke being discharged from a coke oven.
British Steel Corporation

Plate 10 *(below)*
'Highland One', a section of the production platform for BP's Forties oilfield. It is being towed to its site from the construction dock at Nigg Bay on the Cromarty Firth, Scotland, 16 August 1974.
BP Limited

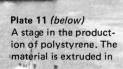

Plate 11 *(below)*
A stage in the production of polystyrene. The material is extruded in viscous strips. When the strips have cooled and hardened they are cut into small granules which are sold to processors to make thermoplastic articles.
BASF

Plate 12 *(below)*
Film-blowing — part of the manufacturing process of polythene film. The polythene is pushed through a ring, air is blown into it to form a balloon 6 or 7 metres high. This is then drawn into rollers which flatten it into two layers of film.
BASF

Plate 13 *(below)*
A research and development laboratory to study and film the behaviour of polymer granules passing through an extruder screw.
BP Chemicals Ltd

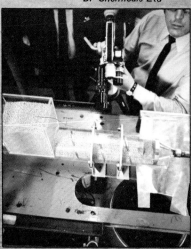

Plate 14 *(top left)*
Sugar beet growing in soil which has too little available manganese. Plants on the left show 'speckled yellows', characteristic of deficiency in this trace element. The foliage of plants on the right has been sprayed with manganese, and they are healthy.
Broom's Barn Experimental Station

Plate 15 *(centre left)*
Six pelleted sugar beet seeds, approximately 5 mm in diameter, showing the structure. Pelleting was first adopted to make seeds round, for precision sowing. It is now also used for dosing seeds accurately with chemicals, either fertilizers or pesticides.
Broom's Barn Experimental Station

Plate 16 *(bottom left)*
This lead glass goblet, with a bucket bowl, has a double-series enamel twist stem. Painted in enamels and gilt, it was made by W. Beilby in Newcastle about 1762. It stands 209 mm high.
Reproduced by permission of the Syndics of the Fitzwilliam Museum

Plate 17 *(top right)*
Built between 1434 and 1446, Tattershall Castle, in Lincolnshire, shows that brick can be as decorative as it is durable.
Edwin Smith

Plate 18 *(bottom right)*
The small-scale manufacture of perchloric acid. Glass containers are used here because perchloric acid and its constituents are corrosive.
Albright and Wilson Ltd.

to a distance and just over a magnetic needle (3 feet from wire ring). Then connected the ends of one of the pieces on A side with battery. Immediately a visible effect on needle. It oscillated and settled at last in original position. On breaking connection of A side with Battery, again a disturbance of the needle.

Made all the wires on A side one coil and sent current from battery through the whole. Effect on needle much stronger than before. The effect on the needle then but a very small part of that which the wire communicating directly with the battery could produce.'

Next Faraday connected a coil of wire to his galvanometer and thrust a bar magnet into the coil (figure 59). He noticed that a current was set up as the magnet moved in, and that a current was set up in the opposite direction when the magnet was drawn out. At first Faraday could only generate short bursts of current. Strenuously, almost fanatically, he sought to produce electricity continuously. Within two months of intensely concentrated work he succeeded in constructing the first dynamo (figures 60 and 61).

Figure 60
A modern 800 megawatt turbine-generator set installed in the Bruce nuclear power station of Ontario Hydro.
C.A. Parsons & Co. Ltd

Figure 61
The apparatus with which Faraday first produced a continuous current of electricity—the first electric dynamo. The copper disc, mounted between poles of the compound magnet, is attached by leads to a galvanometer (in the glass case at the top). When the disc is rotated, there is a continuous deflection of the galvanometer needle. The compound magnet shown here is the original one used by Faraday in his famous experiment; other pieces of apparatus are replicas.
Science Museum, London

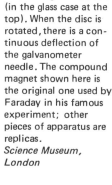

Figure 59
The cylindrical coil and bar magnet with which Faraday showed that magnetism could be used to produce electricity.
Reproduced by permission of The Royal Institution

Genius

What was it about the methods of Faraday which enabled him to outstrip all his contemporaries? Much of his early work was based on his reading of the experiments of others. But he was never satisfied unless he had repeated these experiments for himself; and then his wonderful power of observation usually enabled him to see new facts which had previously escaped notice. As Kohlrausch said, he seemed to 'smell them out'.

Faraday, even from the days of his apprenticeship, made careful notes. In his published *Experimental Researches*, he numbered the paragraphs and welded them together with continual cross-references. His private notes, which are fortunately preserved, show the same method. The last paragraph is numbered 16 041. This same meticulous care is shown in his experiments. In his work on electrochemistry, described in the next section, he first devised, and then thoroughly tested, his method of measuring quantities of electricity; then he showed that the method did not in any way depend on the size of the battery, the number of plates in the battery, or the nature of the connecting wires.

Only then did he proceed to his main investigations. The accuracy of his measurements is quite remarkable and, even with modern equipment, it is difficult to improve on many of them. He anticipated almost all sources of error and possessed phenomenal skill as an experimenter.

He was without mathematical training and he substituted visual models to explain (initially to himself) electrical and magnetic action. Impressed by the beautiful way in which iron filings arrange themselves round a magnet, he pictured 'lines of force' between magnetic poles and between electrified objects, and these lines of force became his working model (figure 62). About them he once wrote: 'I have been so accustomed to employ them and especially in my last researches, that I may have become prejudiced in their favour. However, I have always tried to make experiment the test and the controller of theory and opinion: but neither by that, nor by close cross-examination in principle, have I been made aware of error involved in their use.'

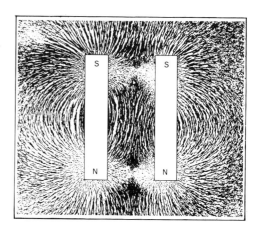

Figure 62
The arrangement of iron filings round a bar magnet, which led Faraday to think of magnetism and electricity in terms of 'lines of force'.
Reproduced by permission of The Royal Institution

Electrolysis

In 1833, at the height of his powers, Faraday directed his attention to the chemical effects of electricity. He first examined the identity of electricity from different sources to show that all electricity (static or current), whatever the source, was of a similar nature. He followed this observation by measuring the quantity of electricity involved in chemical change and the quantity of change produced. He did this by passing electricity through acidified water, collecting and measuring the hydrogen given off, and taking the volume of hydrogen as proportional to the quantity of electricity that had passed. Some of the apparatus he used is shown in the Historical Topics Option. This was the first time such measurements had been made and they led to some far-reaching advances. These are analogous to the advances made in chemistry half a century before when Lavoisier, Black, Cavendish and, later, Dalton and Berzelius, had measured the *quantities of substances* reacting chemically.

From his measurements, Faraday formulated two laws. The first law is:

The quantity of a substance deposited, evolved, or dissolved at an electrode during electrolysis is directly proportional to the quantity of electricity passed through the electrolyte, or in Faraday's own words, 'the chemical decomposing action of a current is constant for a constant quantity of electricity'.

The second law is:

The quantities of different substances deposited, evolved, or dissolved at electrodes by the passage of the same *quantity of electricity are directly proportional to the combining masses of the substances.*

For example, 1 mole of silver (47.0 g) combines with 1 mole of chlorine atoms (35.5 g), and to deposit 1 mole of silver or to evolve 1 mole of chlorine requires the same quantity of electricity, 96 540 coulombs or, as this quantity is sometimes called, 1 'faraday'. But 1 mole of calcium combines with 2 moles of chlorine atoms; to deposit 1 mole of calcium therefore requires 2 faradays of electricity.

In describing electrolysis, Faraday had to look for new names. For these, he consulted William Whewell, afterwards Master of Trinity College, Cambridge, who devised such now-familiar terms as *anode, cathode, electrode,* and *ion*. These terms were based on Greek words. Thus the word *ion* is derived from a word meaning *wanderer*.

Faraday's last years

In 1833 Faraday was elected Fullerian Professor of Chemistry at the Institution and in 1836 was appointed Scientific Advisor to Trinity House, the authority responsible for lighthouses and other sea warning systems.

After his work on electrolysis Faraday became ill and suffered from loss of memory and spells of giddiness. He took a long holiday in Switzerland and on returning to England he rested for four years.

Then in 1845 his health was restored, his mind recovered its former brilliance, and he was able to continue working for another seventeen years.

His last period of work covered many aspects of chemistry and physics. He investigated the reaction (called sulphonation) between naphthalene and sulphuric acid and the supercooling of sulphur; he measured critical temperatures of gases; he determined the chemical nature of rubber; he studied the composition of colloids; and he explained the action of bleaching solutions. He discovered diamagnetism (the tendency of some materials to move within a magnetic field where the field is weakest); he examined thermoelectric effects and produced the first electrochemical series; he explored the conduction of electricity through rarefied gases; and he investigated the regelation of ice (when ice temporarily thaws, then fuses into a solid mass).

He also produced a complete descriptive theory of electricity and of magnetism in terms of 'Lines and tubes of force'. He showed how the capacity of an electrical condenser varies with the insulator between the plates; and he measured the values we call the dielectric constants. Faraday retired from all Royal Institution activities in 1865 to a house at Hampton Court offered to him by Queen Victoria.

Many honours were offered to him, but his Sandemanian creed, to which he remained devoutly attached throughout his life, led him to reject worldly fame. Twice he refused the Presidency of the Royal Society, and he also declined a knighthood. He preferred to remain plain Michael Faraday, scientist, to the end. He died peacefully at Hampton Court in his seventy-seventh year, remembered and commemorated as one of the greatest of all scientists, to be thought of with such names as Newton, Galileo, Lavoisier, and Einstein.

Why do you think that Faraday was such a successful scientist? Which of Faraday's discoveries and inventions do you think was the most important?

The Historical Topics Option provides an opportunity to repeat some of Davy's and Faraday's key experiments in electrochemistry and to study some of their original writings.

Chapter 5
ENERGY AND CHEMICALS FROM COAL, GAS, AND OIL

Most of the energy used to drive machines, to power cars, trains, and ships, and to heat offices and homes is obtained by burning coal, gas, or oil. These fuels are sometimes called fossil fuels because, when they burn, they release chemical energy which was stored up millions of years ago. This stored energy came originally from the sun, but was converted into chemical form by photosynthesis in the leaves of trees and plants. But coal, gas, and oil are not only fuels; they are also a rich source of carbon compounds from which many useful materials such as plastics, dyestuffs, medicines, detergents, and pesticides can be made (figure 63).

THE ORIGINS OF FOSSIL FUELS
The traditional fuel for British Industry since the eighteenth centry has been coal, mined from large deposits in the Midlands, South Wales, the North of England, and Scotland. Nowadays coal is also being mined from deep under the North Sea, although it receives less publicity than North Sea gas and oil. It now seems likely that Britain's fuel needs will be met from her own coal, gas, and oil fields

Figure 63
A chart of energy sources.
BP Educational Service

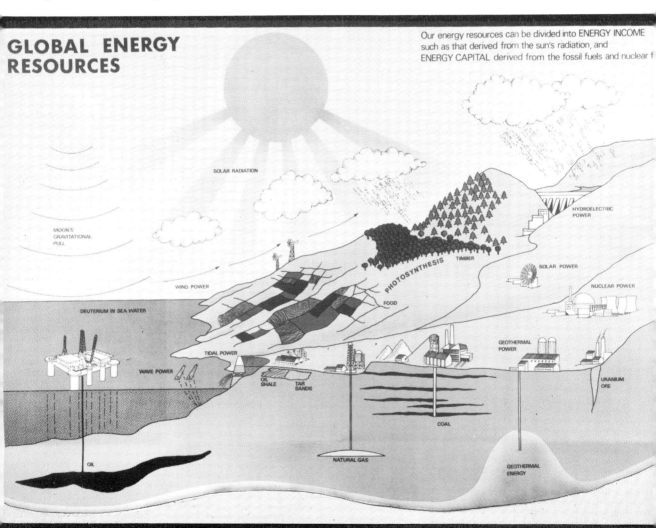

50 Chemists in the World

during the next decade. How is it that large reserves of these three fuels came to be formed under the North Sea?

Compared to the oceans, the North Sea is quite shallow, and geologists see it as a submerged part of the European continent; but it has not been submerged for all of its history. About 300 million years ago the area which is now the North and Midlands of England and the North Sea was a vast tropical swamp in which grew giant reeds, mosses, ferns, and scaly trees. For century after century, for millions of years, these plants grew, reproduced, and died. The dead plants formed a layer of decaying vegetable matter often many feet thick. Sand and mud, washed down by rivers, covered the vegetable sludge. More trees grew on top; they died, decayed, and were in turn covered by sediment. This cycle was repeated many times.

In succeeding ages, the climate and conditions changed, and thick deposits of sediment buried the peat. The weight of these sediments pressed out the moisture and gases from the peat. After millions of years, the sediments over the future coal seams reached depths of over 3000 metres and the heat at these great depths carbonized the peat, converting it to coal. Methane, or natural gas, was one of the products of this carbonization and it is now thought that much of the natural gas under the North Sea, which is composed mainly of methane, was produced when deposits of peat were carbonized to coal (figure 64).

The gas did not stay in the coal seams but escaped upwards into the overlying sediments. These sediments, porous sandstones, were laid down when the area was a desert and the sea was shallow and enclosed. At the same time salt was deposited as the sea evaporated and crystallized. Salt was of great importance, being impervious to gas, because it could seal the top of the porous sandstone, thus trapping the gas that rose from the carbonizing peat. In this way today's coal and gas were formed over 200 million years ago.

Figure 64
An example of fossilized ferns found in coal.
National Coal Board

North Sea oil is younger than the gas and coal: it was probably formed between 100 million and 200 million years ago. This mineral oil is sometimes called petroleum because it is found in rocks (Greek, *petra*, a rock; Latin, *oleum*, oil). Petroleum consists of a complex mixture of many different hydrocarbons. Unlike coal, which can be seen to contain the fossilized remains of the vegetation from which it was formed, petroleum, being a liquid, contains no such obvious evidence.

One theory is that petroleum was produced from the remains of marine animals and plants which sank to the sea bottom to form muddy sediments in places where the water was stagnant and free of oxygen, and where bacteria lived by extracting oxygen from these organic remains, leaving the hydrocarbons which formed oil.

The sediments in which the petroleum was formed are called source rocks. However, the oil did not remain in these rocks, but flowed upwards through overlying sediments until its movement was stopped by an impermeable rock called the cap rock. In the area of the present North Sea, following the formation of the oil, the conditions favoured the decomposition of sediments of chalk, limestone, sandstone, and shale, and these were laid down in great depths. The oil migrated into them, so that they became the main reservoir rocks for petroleum.

Figure 65
A map of Britain's coalfields.
From, Central Office of Information (1975) Energy. C.O.I. Reference Pamphlet 124. H.M.S.O.

THE COAL INDUSTRY

Mining for coal

There are in Great Britain about 240 working collieries employing roughly 240 000 miners to produce around 120 million tonnes of coal each year (figure 65). These figures include open-cast mining but most of the coal is extracted deep underground by machines which automatically cut the coal and load it onto conveyor belts. Many of these machines are guided by a γ-ray sensing device which steers the machine along the coal seam and prevents it from cutting into the floor or the roof. The roof of the seam is held up by hydraulically powered supports which can be moved forward and reset against the roof by remote control. The coal is carried by trains of minecars, or by conveyor belt, to the bottom of the mineshaft to be brought to the surface (figure 66).

Mining for coal has always been dangerous because of the possibility of fires, explosions, flooding, and roof falls. In the nineteenth century, accidents were more frequent than they are today. The mines were poorly ventilated and methane escaping from the coal seams caused explosions when allowed to accumulate. As mentioned in chapter 4, Humphry Davy contributed to the development of the safety lamp which helped miners to avoid the danger of explosions (plate 6).

Figure 68
Fuel share of the U.K. energy market on the basis of heat supplied to final users.
From, British Gas Corporation. Annual Report and Accounts 1975–76.

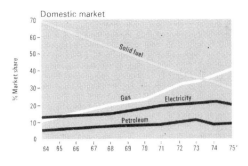
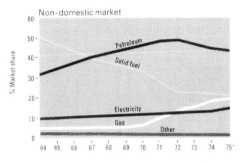
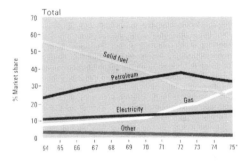

Figure 66
A coal-cutting machine at Cotgrave colliery, South Nottinghamshire area, fitted with a nucleonic probe which automatically steers the machine to keep it within the coal seam.
National Coal Board

Figure 67
Taking continuous samples of the atmosphere in the mine: a junction box of a four-tube bundle at Harworth colliery. One of these junctions is needed about every 500 m underground.
National Coal Board

Today scientists are still working to improve the safety of mines. Fires may result from the spontaneous combustion of a coal seam which becomes overheated by absorbing oxygen. We now know that the gases given off the coal can be used to give advance warning of a dangerous rise in temperature. The amount of gas involved is small, so that sensitive physical methods of analysis, such as chromatography and spectroscopy, have to be used. In some mines, samples of the mine atmosphere are continuously fed, through fine tubes running for several kilometres underground, to instruments on the surface for analysis (figure 67).

How coal is used

Coal is used in three main ways: as a fuel to be burned, as coke, and as a source of chemicals (figure 68). Just over half the coal mined in Britain is burned in power stations to generate electricity (plate 8). In 1976/77 77.8 million tonnes of coal were used to fuel power stations. This accounts for 63 per cent of total coal consumption, and another 37 per cent was consumed by industrial and domestic users. In addition 1.4 million tonnes were exported, mainly to EEC countries.

In power stations the coal is ground to a fine powder and injected into the boilers through

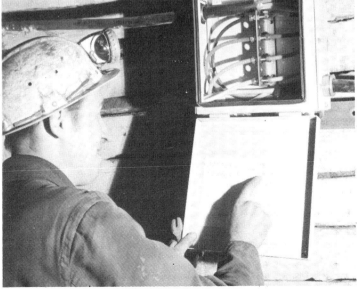

Figure 69
Coal mine and power station working in harness: the Lee Hall Colliery and Rugeley Power Station.
National Coal Board

wide nozzles where it burns like a gas flame. Coal is expensive to transport and so, whenever possible, the need to carry coal across country is avoided by building power stations alongside collieries so that the coal can be conveyed direct from the pit head to the power station (figure 69). When coal is burned properly and completely, as in industrial boilers, there need be no smoke.

However, it is difficult in most fireplaces and household boilers to burn coal without producing smoke unless special care is taken. Recently domestic boilers and room heaters have been designed which will burn cheap coal efficiently. More often, the coal is specially processed to make a smokeless fuel which can be burned in an open fire. Because of these developments it has been possible to extend smokeless zones in large cities.

When coal is heated strongly in the absence of air to a temperature of about 1000 °C it splits into three parts:
coal gas consisting of hydrogen, methane, and carbon dioxide,
several liquids such as benzole and tar,
and solid coke which is a form of carbon.

This process used to be carried out to provide gas for domestic heating and lighting; the coke produced was then available in large quantities as a by-product for use as a smokeless fuel. Now coke is made in special ovens. These are tall, thin chambers, and thirty or more are set side by side, interleaved with spaces through which white-hot burning gas can pass to heat the coal. Crushed coal is put into the ovens from the top and is then left to heat up. One or two days later, when all the gases and liquids have been driven off, the doors at the front and back of the oven are taken off and a large ram pushes out about fifteen tonnes of glowing coke into a wagon, called a coke car (plate 9). The coke car runs on rails to a tower which has water sprays to cool the coke and prevent it burning in the air.

Coke is used in blast furnaces to extract iron from its ores; it is also an important industrial fuel in foundries where metal castings are made, a source of carbon in the chemical industry, and a domestic fuel.

About two-thirds of a tonne of coke are made from a tonne of coal, the remainder being driven off as gas and volatile liquids. Now that the domestic gas industry has been converted to natural gas, the coal gas from the coke ovens cannot be distributed to domestic customers; instead it is sold to nearby industries such as steel, glass, and chemical works. The volatile liquids are condensed to give crude tar, benzole, and ammonia. The *tar* is processed to obtain road tar, waterproof roofing materials, disinfectants, liquid fuels, dyestuffs, and many other products. The *benzole* is a source of chemicals used to make plastics, dyestuffs, and solvents. Most of the *ammonia* is converted to ammonium sulphate which is used as a fertilizer (figure 70).

Figure 70
Some by-products of coal.
National Coal Board

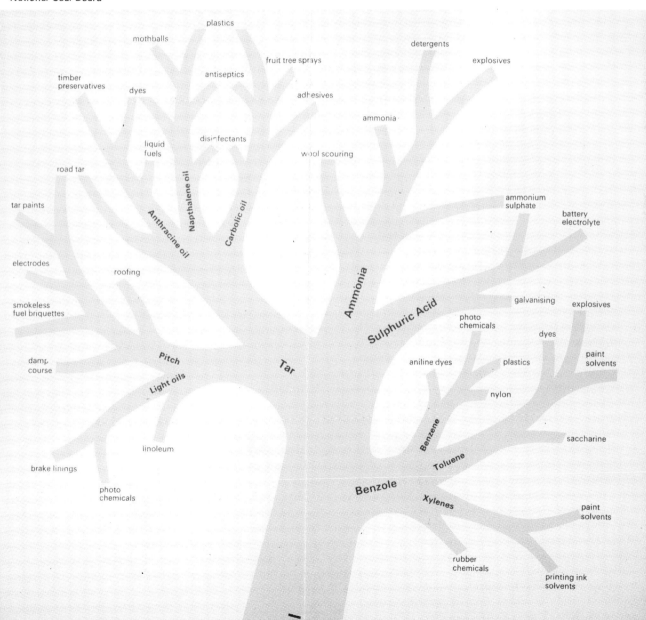

THE GAS INDUSTRY

From coal to natural gas

In recent years there has been a series of changes in the structure of the British gas industry which illustrates the way in which an industry must keep modifying its methods to remain as competitive as possible. Before 1960 the gas industry was entirely a manufacturing industry. Every major town had its gas works where coal was carbonized by being heated strongly in the absence of air to drive off the coal gas. After purification, this gas was distributed locally, mainly to domestic consumers. High-grade coal had to be used, so the gas was expensive in comparison with other fuels, and little was used by industry.

By 1960 many new oil fields had been discovered. The result was a world surplus of oil. Coal gas then became very expensive compared with other fuels, and compared with electricity which was generated using cheap oil or low-grade coal. In looking for cheaper gas, the industry developed techniques for transporting liquefied natural gas by ship from the oil fields in Algeria and Libya to a terminal at Canvey Island. This was the first gas to be distributed nationally rather than locally, by laying an underground pipeline across Britain.

At about the same time, a new process, the Lurgi process, was devised to make gas from low-grade coal. Another process was invented to make gas, now called town gas, from the lighter (naphtha) fractions obtained when crude oil is fractionally distilled. By 1969, 112 plants for manufacturing town gas from petroleum had been built.

But most important of all was the discovery of reserves of natural gas under the North Sea. The first strike was made in 1965 and since then many more gas fields have been found. Natural gas requires only minor treatment before it is distributed by pipeline from the gas fields to consumers. It is therefore a cheap fuel and so the gas industry decided to convert its whole operation to one of distributing and selling natural gas.

Figure 71
Construction of the 34-mile, 30-inch marine pipeline to carry gas from the Leman field in the North Sea. The pipeline is brought ashore at Bacton, Norfolk. The pipelaying barge is visible in the background.

All gas appliances had been converted to burn natural gas by the Autumn of 1977.

Looking to the future, it is clear that the supplies of natural gas are limited. By the time they begin to run out oil is also likely to be in short supply. The only fossil fuel available in large quantities is coal, and so eventually the gas industry will probably have to make its gas from coal again — by an up-to-date version of the Lurgi process. Another possibility is to obtain gas from coal while the coal is underground — which would obviously save the trouble of mining it first.

Figure 72
A map of Britain's gas grid and North Sea gasfields.
From, Central Office of Information (1975) Energy. C.O.I. Reference Pamphlet 124. H.M.S.O.

―――――――――――――――――

The social, geographical, and economic factors which influence development of manufacturing industries are discussed in Chapter 10, *Handbook for pupils*, and Option 8.

―――――――――――――――――

Distributing natural gas

Gas from the North Sea fields is brought ashore by pipeline to terminals on the east coast (figure 71). At the terminal, liquids are removed, the gas is dried and measured, and its pressure is regulated before it passes from the producing company to a neighbouring terminal belonging to the British Gas Corporation. Here the gas is filtered, blended with gas from different fields so that the mixture will produce the guaranteed minimum amount of heat when burned, and finally given a characteristic smell by the addition of traces of a special chemical. The gas then

Chemists in the World

enters the national grid of large-diameter, high pressure, underground pipelines (figure 72).

The flow of gas through the grid is controlled from a centre at Hinckley, in Leicestershire. The controllers must make sure that the demand for gas is met and that the required gas pressure is maintained at all outlets. The flows and pressures of the gas throughout the grid are recorded by automatic sensors and transmitted by telephone and radio links to a computer at the control centre. The computer processes the information and presents a summary to the engineers on a visual display panel. The engineers use weather forecasts to help predict the demand for gas, and, again with the computer, they determine the settings of the remote-controlled valves used to regulate the gas flows.

How gas is used

Most natural gas is burned to produce heat energy for three types of customer: domestic, commercial, and industrial (figure 73). The

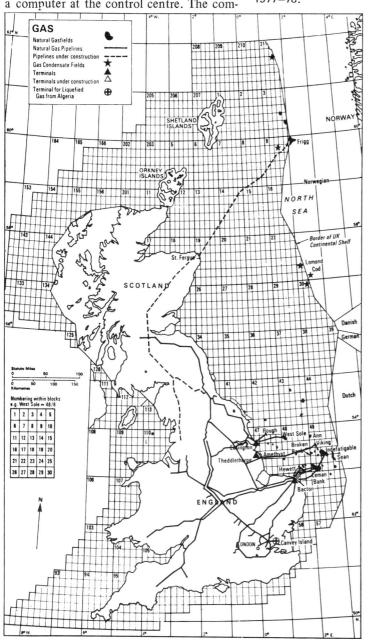

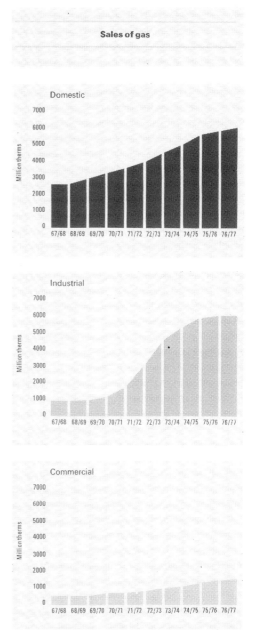

Figure 73
Sales of gas in Britain.
From British Gas Corporation. Annual Report and Accounts 1977–78.

Energy and chemicals from coal, gas, and oil 57

gas used domestically and commercially is mainly for heating and cooking in people's homes and places of work. Before the discovery of natural gas, the gas from coal and oil was too expensive to be used much in industry. It was only bought by industries requiring a high-quality, easily controlled fuel particularly suitable for applications such as heating glass furnaces and pottery kilns, and for food processing and metal finishing.

Since the discovery of natural gas, much more gas has been used in industry, for example in paper making, textile and leather processing, and for making vehicles. The chemical industry now uses a lot of natural gas, but not for burning; the gas is used to make hydrogen, in turn used in the manufacture of ammonia for fertilizers.

Little natural gas is used to generate electricity, but in order to even out the peaks in the demand for gas, the industry has contracted to supply gas on an interruptible basis to some users. One of these contracts is with the Central Electricity Generating Board, which has converted two power stations to burn coal or natural gas, but gas is only used when a surplus is available.

THE OIL INDUSTRY

Drilling for oil

Crude oil is found almost everywhere. The richest producing areas at present are the Middle East, the United States, Latin America, Africa, and the Soviet Union. But crude oil is also found in many other countries including Borneo, the Netherlands, and Japan. In recent years, organizations have started looking for oil in offshore areas, and have discovered important reserves under the North Sea and in Alaska (plate 10). Crude oil is usually a thick, dark liquid looking like black treacle. But the proportions of the different hydrocarbons in the oil vary greatly from one area to another.

The decision where to drill for oil is usually left to geologists who are experts in the study of rocks. Many geologists are continually searching the earth, often in remote places,

An oil well

Figure 74
Diagram of an oil well.
Shell

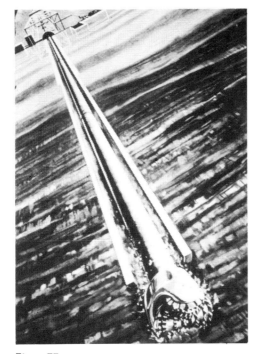

- crown block
- platform
- cable
- drill pipes
- travelling block and hook
- rotary hose
- swivel head
- kelly
- derrick floor
- rotary table
- engines
- blow-out preventers
- vibrating screen
- mud ditch
- slush pump
- mud suction pit

for rock formations in which they believe oil may have accumulated (figure 74). Modern drills bite down into the rock with a circular motion, rather like a brace and bit boring a hole into wood. The drill consists of a bit or cutting piece attached to a rotating column of pipe (figure 75). The pipe is turned by an engine at the surface. Also at the surface is the derrick, which looks rather like an electric pylon, from which the lengths of pipe are fed into the drill hole. The bit is kept cool by a current of lubricating mud which flushes out the drill cuttings. This mud, together with valves called blow-out preventers, helps to prevent a blow-out of oil. A blow-out, as well as being wasteful, can wreck the derrick and set the oil well on fire. With modern methods, holes as deep as eight kilometres have been drilled. Also, using off-shore platforms, drilling has been extended to areas under the sea (figure 76).

Figure 75
Drilling in progress. Artist's impression of a drilling bit, drill string and casing in strata.
Shell

Figure 76
Three types of platform in the North Sea: *a* and *c* are permanent installations.
a Amoco–A, a gas production platform on the Leman Bank field.
British Gas Corporation
b A self-propelled semi-submersible drilling platform, used for oil exploration.
c The giant Brent 'B' production platform. The platform stands in 140 m of water, with the top of the drilling mast 108 m above the water.
b and c Shell

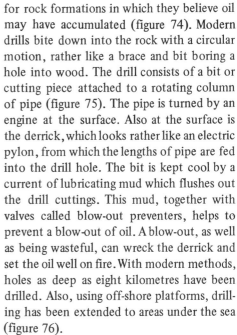

Energy and chemicals from coal, gas, and oil

Figure 77
Tugs berthing the oil tanker *Esso Demetia* at Milford Haven. Built in 1973, she has a deadweight tonnage of 250 000, an overall length of 340.46 m, a moulded breadth of 51.8 m, and a fully loaded draught of 20.117 m.
Esso

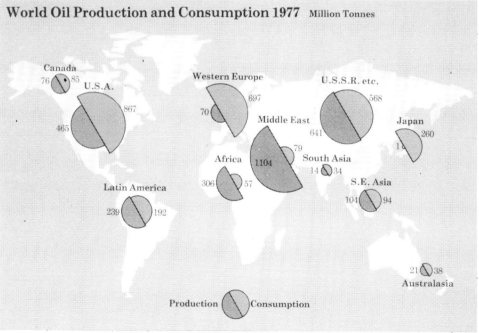

Figure 78
World oil production and consumption 1977, shown in million tonnes.
From The British Petroleum Co. Ltd (1978)
BP statistical review of the world oil industry 1977

When the drill strikes oil, the underground pressure is likely to force the oil up the bore hole and may continue to do so for many years; but if the pressure is not sufficient, or if it falls off, the oil must be pumped. On average, only about 35 to 45 per cent of the oil below ground can be brought to the surface. New techniques are helping to improve this rate of recovery.

Transporting the oil

Although oil is found in most parts of the world, many of the richest fields are thousands of kilometres from the industrial centres where oil is most in demand. Getting the oil to where it is wanted involves two journeys. First the crude oil has to be transported from the oil field to the refinery; then the refined oil products have to be transported from the refinery to the markets where they will be sold. Sometimes refineries are built near the oil fields. More often they are built in the consuming areas (figure 78).

The oil starts its journey from the wells by pipeline, which may take it as far as the refinery. But transport by pipeline is usually more expensive than transport by water, and wherever possible ships called tankers are used for moving oil in bulk. Over one-third of the world's shipping tonnage is made up of oil tankers, and some of these are among the biggest ships afloat (figure 77).

- -

Crude oil and refinery products are transported by road, rail, ship, and pipeline. What are the dangers involved in transporting these materials? What are the advantages and disadvantages of the different methods of transport (figure 79)?

- -

Refining the oil

A modern oil refinery, with its gleaming metal columns, its miles of intricate piping, and its automatic control panels, is a complex structure designed to convert crude oil into a vast range of useful products.

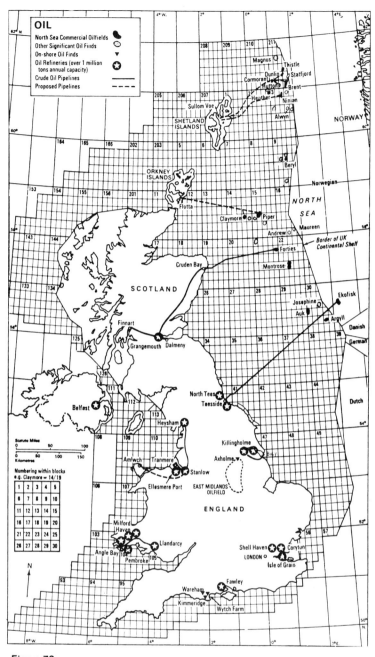

Figure 79
A map of Britain's oil fields and oil refineries.
From, Central Office of Information (1975) Energy. C.O.I. Reference Pamphlet 124. H.M.S.O.

Surprisingly, the oil being processed is never seen as it flows invisibly through the pipes from one process to the next. Crude oil, which is a mixture of thousands of different hydrocarbons, is of little use as it comes from the ground. At the refinery it is treated in three main stages to provide the substances required by the customers of the oil company. The three stages are *separation, conversion,* and *purification.*

Energy and chemicals from coal, gas, and oil 61

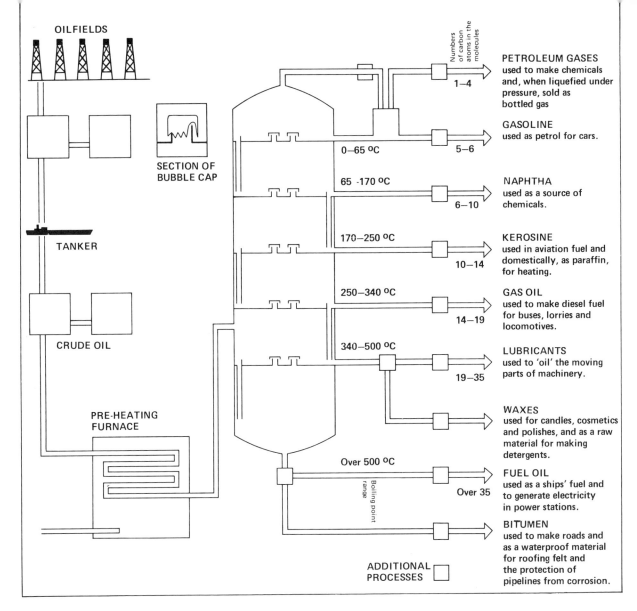

Figure 80
Diagram of a fractionating column in an oil refinery. Vaporized crude oil passes up the column, and separated fractions, as they condense on the trays, are collected off the side. Bubble caps (on each tray there may be many hundreds) help force the rising vapours through the condensed liquids, and so ensure precise separation. The part of the crude oil which does not vaporize is drawn off as a liquid from the base of the column.

The first stage is to separate the hydrocarbon mixture into 'fractions' of similar molecular size. This separation is carried out by distilling the crude oil. Distillation in an oil refinery is essentially similar to the laboratory process except that instead of boiling off the different fractions one after the other, nearly all the crude oil is vaporized at once and the separate fractions are collected as the vapours condense back into liquids (figure 80). The vaporized crude oil is passed into a fractionating column (which may be as high as sixty metres), and the hydrocarbon fractions, boiling at different temperatures and therefore at different heights in the column, are separated and collected as they condense.

The second stage in the refining process is the most tricky and the most interesting. It is necessary because some oil fractions are needed more than others. Over a period of years, the demand for any particular product can vary a great deal.

Take the example of kerosine (paraffin). At the beginning of this century, all that was extracted from crude oil was the kerosine fraction which was used for lighting in paraffin lamps; the rest of the oil was wasted. With the invention of the internal combustion engine, the demand for the gasoline (petrol) fraction grew while the demand for kerosine fell because people started using electricity for lighting. Then in the mid-forties the demand for kerosine increased again because it was the main ingredient in the fuel for jet engines and rockets.

In a modern refinery unwanted fractions are no longer wasted but are converted into the products most in demand. On the one hand, large molecules of a heavy fraction such as gas oil can be split to give two small molecules of a lighter fraction such as gasoline. On the other hand, two or more molecules of a light fraction can be joined together (polymerized) to give a larger molecule of a heavier fraction.

Breaking down larger molecules into smaller ones is called *cracking* and it is one of the main conversion processes used in an oil refinery. Originally, cracking was carried out by heating the heavier oil fractions under pressure: this process is known as thermal cracking. Nowadays it is more usual to crack the oil in the presence of catalysts, for example alumina or silica, because in this way the process can be carried out at much lower temperatures and pressures. As well as breaking down the large molecules, another effect of catalytic cracking is to rearrange the atoms in the molecules: straight chains of carbon atoms may become branched chains or rings. When making gasoline this conversion is desirable because the branched chain and ring hydrocarbons greatly improve the efficiency of the fuel in motor-car engines. The modern internal combustion engine requires petrol of a very high performance. So, in addition to cracking, many refineries operate a *reforming* process to improve the quality of the gasoline fraction which is to be used as petrol.

Using conversion processes of this kind, the oil refiner has become a chemical conjuror able, within limits, to draw out of the hat the required fraction with the required properties in the required amount.

The final stage of refining, purification, involves the removal of impurities from the different fractions to give finished oil products. Some of the commonest impurities are sulphur compounds which, unless they are removed, give off poisonous fumes of sulphur dioxide when the oil is burned (figure 81). The amount of sulphur present depends on the source of the original crude oil.

How many of the products of refining crude oil can you find in your home? How and where are they stored? What are they used for?

Chemicals from oil
Only about 5 per cent of the oil extracted from the ground is used to manufacture chemicals, but from this 5 per cent comes more than three-quarters of the world's supply of carbon (organic) compounds. Most of the chemicals are made from the naphtha fraction from the distilling column which is cracked to form reactive hydrocarbon gases such as ethene (ethylene) and propene (propylene). These simple substances are the

Figure 81
The Shell flue gas desulphurization plant at Showa in Japan. The process removes 90 % of the sulphur oxides from stack gases, and thus permits the use of low quality fuels which would otherwise cause too much air pollution.
Shell

Energy and chemicals from coal, gas, and oil

building blocks used to make products such as solvents, detergents, industrial chemicals, agricultural chemicals, plastics, and synthetic rubbers (figure 82).

Oil has sometimes been called 'liquid gold'. Do you agree that this is a good nickname? You can read about the work of the men who run a large pyrolitic cracking plant in a petro-chemicals factory in chapter 10 of the *Handbook for pupils*.

FOSSIL FUELS AND THE ENVIRONMENT

Fossil fuels are valued as sources of energy and chemicals, but extracting these fuels from the earth, and transporting and burning them, may damage the environment in a number of serious ways.

When coal is mined, waste rock is deposited on the surface in tips which often become a permanent and unattractive feature of the landscape (figure 83). These tips may be a danger to life, for if they are not well drained, water will accumulate in them and they may suddenly slip and pour down in a rush of

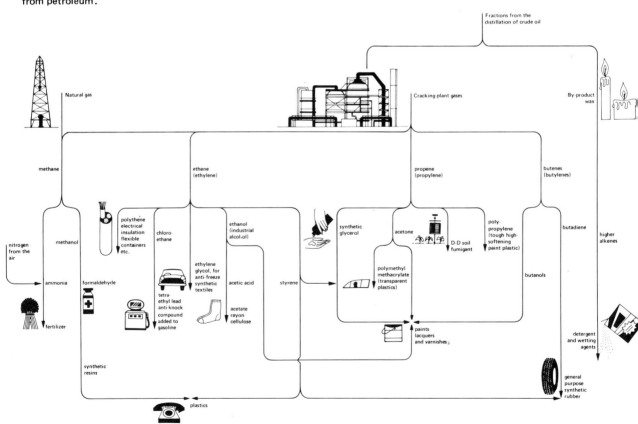

Figure 82
Some of the chemicals and chemical products that can be obtained from petroleum.

watery mud, as happened at Aberfan in South Wales in 1966. Another consequence of mining is subsidence of the land which damages buildings and roads (figure 84).

The extensive drilling for oil under the sea is a new hazard because an accident at a well can lead to a large amount of oil gushing out into the water where it is lethal to many marine organisms. Oil tankers are another source of pollution of the oceans. These ships use sea water to wash out their tanks, and some oil escapes into the water despite operating procedures designed to prevent this. Accidents may also occur when the oil is

Figure 83
a An open-cast coal mine.
Hazel Geoffrey
b Landscape restored after open-cast mining.
The Forestry Commission

Figure 84
Severe damage from a tensile strain of 4.3 mm/m in a block of houses 26 m long. The fracture widens appreciably towards the top. This damage was the result of extracting a seam 1.7 m thick at a depth of only 137 metres.
From Mining Department (1975) Subsidence Engineers' Handbook. *National Coal Board*

Energy and chemicals from coal, gas, and oil

being loaded on or off the tanker at the terminal, and large amounts of oil can be released if a tanker is involved in a collision or is wrecked. A well-known example of massive pollution by oil followed the grounding of the *Torrey Canyon* off Land's End in 1967, when 120 000 tonnes of crude oil were spilled. Unfortunately many of the detergents used to disperse the polluting oil are also toxic. (Figure 85.)

Burning huge tonnages of fossil fuels clearly produces large amounts of heat. It also produces carbon dioxide together with smaller quantities of pollutants such as sulphur dioxide, nitrogen oxides, carbon monoxide unburned hydrocarbons, and solid particles. It is estimated that the carbon dioxide content of the atmosphere may increase by 25 per cent between now and the end of the century, and there is speculation that this could lead to a small increase in the mean surface temperature of the earth and a significant effect on the world's climate, because carbon dioxide absorbs and traps heat radiation escaping from the atmosphere. However, as the use of fossil fuels declines, the level of carbon dioxide in the air will also decline because much of it will dissolve into the seas and some will be taken up by photosynthesis, as the growth of plants is stimulated by the higher level of carbon dioxide in the air.

All the energy generated by burning fossil fuels eventually ends up as waste heat. The total amount of heat involved is very small compared with the energy which comes from the sun, so on a world scale the release of heat is probably insignificant, but the use of fossil fuels is concentrated in densely populated industrial areas and, particularly near power stations, the waste heat can have a marked effect on the local environment. For example, water extracted from rivers for use as a coolant in power stations is returned to the river at a higher temperature, and this can upset the balance of plant and animal life in the river.

Most coal and oil contains sulphur compounds which form sulphur dioxide when they are burned. In a moist atmosphere, sulphur dioxide forms sulphuric acid which is corrosive and which, when carried into the soil by rain, may reduce the fertility of the land by making it too acid. In large towns and cities, the atmosphere is also polluted by carbon monoxide, hydrocarbons and nitrogen oxides released from the exhausts from the engines of motor vehicles. It is clearly desirable to reduce this pollution which is a danger to health, and legislation in many countries is setting new standards and forcing car manufacturers to install devices which purify exhaust gases. Unfortunately these measures will lead to an increase in fuel consumption just when we are trying to save fuel.

Figure 85
A fire-fighting vessel spurting water at the 'Bravo' platform, in the Norwegian Ekofisk oilfield, after a blow-out of oil and gas on April 22, 1977.
Associated Press Ltd

———————————————

There is often a conflict between the desire to reduce pollution and the need to be economic and conserve energy. Who should decide which is the more important objective, and what factors should they consider when taking decisions (figure 86)?

———————————————

FOSSIL FUELS IN THE FUTURE

It may seem wasteful to burn fossil fuels in vast quantities for energy when they are such a valuable source of chemicals, but only a very small percentage of the coal, gas, and oil extracted from the ground is now required for the manufacture of new materials. The world's demand for energy is doubling approximately once every ten years, and at present this demand can only be met by burning more and more coal, gas, and oil which together supply about 90 per cent of the energy used.

But these fuels which took millions of years to form cannot be replaced. On current estimates the production of oil and gas will reach a peak towards the end of this century, and most reserves will be exhausted by the year 2100. Such estimates are very uncertain because large new oil fields may be discovered, and new techniques may be developed either for extracting more of the oil from existing fields or for obtaining oil from the huge reserves in tar sands and oil shales. Coal deposits are very much larger and are expected to last for 150 to 200 years, even if coal consumption continues to increase at the same rate as it has in recent years.

If the increasing demand for energy is to be met in the future, two different types of problem will have to be solved. On the one hand, new methods for generating usable forms of energy will have to be devised and proved on a large scale; on the other hand, more efficient techniques for using and conserving energy must be put into practice.

Probably the most important alternative to fossil fuels for generating electricity is nuclear power, which may one day be used to produce all electricity. Other sources of energy which can be harnessed are winds, tides, running water, direct sunlight, and the heat stored underground in volcanic regions. The opportunities for the better use and conservation of energy are considerable because it has been estimated that at the moment half the energy produced by burning fossil fuels is wasted. Some of the waste occurs when the heat is converted into a more convenient form of energy such as electricity; some energy is lost in carrying energy from where it is produced to where it is used; and yet more energy is wasted because machines are used inefficiently, and buildings are poorly insulated against heat losses.

Figure 86
The sour water stripping facilities at Fawley. These remove hydrogen sulphide and ammonia from the refinery's effluent water to minimize pollution, as required by law.
Esso

Some say that coal, oil, gas, and electricity are supplied to our homes so cheaply that we can afford to waste them. Can you think of ways in which you could conserve energy in your home? Would you be more careful to conserve energy if it were more expensive?

It has been estimated that the solar energy falling on the surface of the earth is about 5000 times the predicted world energy requirement in the year 2000. If so, why is there talk of an energy crisis? Would you like to spend your life as a scientist developing new methods of harnessing the sun's energy?

Since the world's reserves of coal, gas, and oil are limited, should their use be controlled so that each fuel is used only for those purposes that use it most efficiently? Which fossil fuel do you think should be used for: domestic heating; chemicals manufacture; road, rail, and air transport; and the generation of electricity?

Chapter 6
POLYMERS FROM PETROLEUM

The common names for plastics, and the chemicals used to make them, differ from the systematic, theoretical names preferred by chemists. In this chapter common names are used but the table gives systematic names.

Common name	Systematic name
Ethylene	Ethene
Polythene	Poly(ethene)
Hexamethylene-diamine	Hexane–1,6–diamine
Adipic acid	Hexanedioic acid
Ethylene glycol	Ethane–1,2–diol
Terephthalic acid	Benzene–1,4–dicarboxylic acid
Vinyl chloride	Chloroethene
P.V.C.	Poly(chloroethene)
Formaldehyde	Methanal

POLYMERS FROM PETROLEUM

Polymerization

Substances with very large molecules are called *polymers*. The word polymer derives from two Greek words meaning many units, and is an apt description of the large molecules which are formed by the linking together of smaller, often identical, molecular units. There are many different types of polymer. Some occur in nature – for example, proteins, starch, wool, silk, cotton (cellulose), and rubber. Others are man-made – plastics, synthetic fibres, and synthetic rubbers, for example.

Plastics are polymers which are solid at ordinary temperatures, but at some stage in their manufacture are made capable of being easily shaped, or plastic. (This is the origin of the name 'plastics'.) Plastic materials are usually shaped into useful articles by heat, or pressure, or a combination of the two (plate 11).

Although plastics are now common, many people do not know what they are or how to distinguish one from another. Often it is thought that plastics are just one material called 'plastic', but this is quite wrong. The plastics are really a group of materials and their properties vary greatly. Some are soft and flexible; others are hard and can be worked as if they were wood or metal (figures 87 and 88). Some are especially useful because they resist acids and other chemicals.

Some plastic polymers, such as nylon and terylene, can be specially processed to produce strong threads which can be used to

Figure 87
Thermoset moulding material is used for a variety of car engine components.
Bakelite U.K. Ltd

Figure 88
These worktops covered with decorative laminate – heat-resistant and easy to clean – are another example of a thermosetting plastic.
Perstorp Warerite Ltd

make fabrics. In this form the polymers are called fibres (figure 89).

In the plastics industry there are two chief methods of making large molecules. The object is to use a chemical reaction called polymerization so as to make a large number of small molecules, called monomers, link together to form a large one. The processes for polythene and nylon illustrate the two methods.

The first method of polymerization

Polythene (figure 90), is a white waxy solid obtained by polymerizing ethylene gas (C_2H_4). (This gas is obtained by cracking petroleum.) When heated at 100–300 °C under pressures several thousand times the pressure of the atmosphere, the molecules of ethylene gas join to form long-chain molecules. As in many polymerization reactions, the presence of a small amount of catalyst makes the reaction very much faster. About 1000 or even 10 000 molecules may link together to form a long-chain molecule.

The idea that very long molecules could result from the chemical linking of many identical small molecules, a process described as *addition polymerization*, was first suggested by the German chemist Hermann Staudinger, for many years professor of chemistry in the University of Freiburg. He began his work on large molecules about 1922 and first used the term 'macro-molecules'. He was the founder of polymer chemistry.

Figure 91
This bottle is made of PVC. It is shatterproof and light.
ICI Ltd, Plastics Division

Figure 89
The 100 000 tonne super-tanker, *British Admiral*, carries nearly five tonnes of terylene rope. Each rope has a minimum breaking strength of 57 tonnes. They are used during mooring operations.
ICI Ltd, Fibres Division

Figure 90
Polyethylene film used for packaging loaves of bread.
Bakelite Xylonite Ltd

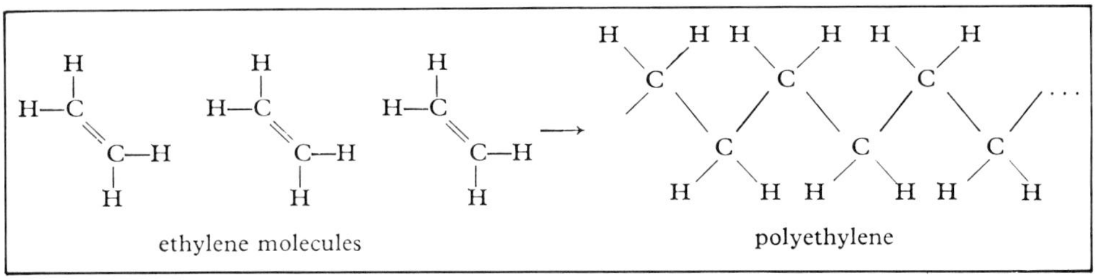

Figure 92
Linking together of ethylene molecules to form polythene.

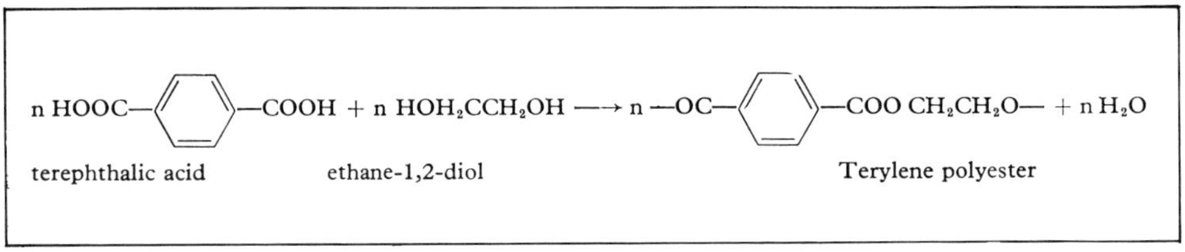

Figure 93
Linking together of terephthalic acid and ethane–1,2–diol molecules to form terylene.

Figure 94
a Random arrangement of molecular chains of different lengths with the chains continuously coiling and uncoiling. The forces between molecules are weak and the polymers, which are soft and flexible, are classified as rubbers.
b Forces between the chains are stronger and the chains less flexible to give an amorphous structure typical of many thermoplastics.
c Cross-linking between the chains creates, in effect, a single giant molecule — as in thermo-setting plastics such as Bakelite.
d Several chains lie parallel to one another for all or part of their length to produce a close-packed orderly arrangement — as occurs in a number of thermoplastics. These regions of order are described as being crystalline, and the forces between the chains are much stronger.
e Stretching of crystalline polymers results in further alignment of the molecular chains and increased strength. Nylon and Terylene filaments are stretched to convert them into textile fibres.
f In some polymers, the molecular chains are branched and the branches may crystallize among themselves.

Many of the most important industrial plastics — for example, PVC (figure 91) and perspex, as well as polythene (plate 12) — are made from simple chemical substances by addition polymerization. The simple substances that can polymerize in this way have one property in common: there is a double-bond between two carbon atoms; these are called *unsaturated molecules* (figure 92).

Do you prefer to drink out of a metal, glass, or plastic cup? Which type of cup is most commonly used in the home, at camp, in automatic vending maching, or in a canteen?

The second method of polymerization

In this method, small molecules that have a mutually reactive group at each end are linked together, as in the reaction for making nylon (figure 93). This was discovered by the American chemist W.H. Carothers, another famous pioneer of polymer chemistry. His method depended on the reaction of organic bases (amines) with organic acids. Amines contain the reactive amine group $-NH_2$, and organic acids contain the reactive carboxyl group $-CO_2H$. When a mixture of an amine and an organic acid is heated, the two groups react: water is split off and amine and acid join to form a larger molecule: an amide.

The key to Carothers's discovery was that he used reactants each of which contained two reactive groups — a diamine and a dicarboxylic acid. These can be represented by the general formulae:

H_2NRNH_2 $HO_2CR'CO_2H$
diamine   dicarboxylic acid

where R and R' are shorthand symbols for the hydrocarbon part of the molecule between the reactive groups.

When a mixture of these two reactants is heated, they combine in the normal way:

$$H_2NRNH_2 + HO_2CR'CO_2H$$
$$\downarrow$$
$$H_2NRNHOCR'CO_2H + H_2O$$

The interesting point here is that the product still contains two reactive groups — an amine group at one end and a carboxyl group at the other. Thus the first product can react further — with more acid at one end and with more amine at the other. When this happens, the product still has reactive groups at each end and can therefore continue reacting. The reaction goes on and on. The chain formed gets longer and longer. Eventually a very extended long-chain molecule results, called, in this example, a *polyamide*. It contains a large number of the following chemical units

–HN R NH CO R'CO–

all joined end to end.

Usually, nylon is the polyamide made by

Chemists in the World

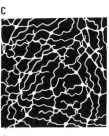
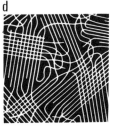

starting with hexamethylene diamine ($H_2N(CH_2)_6NH_2$) and adipic acid ($HO_2C(CH_2)_4CO_2H$). It is often called nylon 66, which is a short way of showing that the diamine and the diacid used to make it each contain six carbon atoms in the molecules. Nylon 66 was the most important of several polyamides discovered by Carothers during his researches, which began in 1928. He died nine years later at the age of forty-one, but not before it was seen that nylon was a commercial success. As a yarn for making stockings it entered the American market in 1940. Shortly afterwards, because of its toughness, nylon also began to be used by the plastics industry.

The reaction of two molecules to give a larger molecule, with the splitting off of a small molecule (usually water), is called a *condensation reaction*. The use of this kind of reaction to build up polymers (as in the nylon process) is referred to as *polycondensation*. A number of plastics and synthetic fibres are made in this way. Starting, for example, with ethylene glycol (also used by the motorist as an anti-freeze agent for his radiator) and terephthalic acid, the polymer obtained is a polyester called polyethylene terephthalate — better known by its commercial name, Terylene. The large Terylene molecules are built up in a way similar to the nylon molecules — although here carboxyl groups are reacting with hydroxyl groups (–OH) instead of amines.

Terylene was discovered in Britain in 1941 by Whinfield and Dickson and is now produced on a very large scale, chiefly for use as a fibre.

THE STRUCTURE OF POLYMERS

We have pictured polymer molecules as being made up of very long chains of simple chemical units, usually 1000 or more of them in an actual sample. If we could take a photograph of the polymer, the picture might look like a tangled ball of many long pieces of string. To get a more realistic picture we would have to use a motion camera, because the molecules of a polymer, like all chemical molecules, are in a state of motion. The molecular chains possess kinetic energy, and are vibrating and rotating and sliding over one another, rather like a lot of live eels in a bucket (figure 94).

In polymers the molecular movement tends to be hindered by the entanglement of the long chains with one another. But there is also another restraining influence on the movement of the molecules, from attractive forces between the molecules. These forces are much weaker than the chemical forces which bind atoms together to form molecules, and they are really effective only when the molecules are very close together. For example, it is because of these forces between molecules that, under appropriate conditions, gases condense to liquids and liquids change to solids. The existence of forces of attraction between long-chain molecules is one of the important factors that influence the properties of polymer materials.

When a polymer like polythene is heated, the molecules gain energy and move about more vigorously. The chains therefore become further separated and the attractive forces are weakened. As a result the polythene becomes softer and more flexible, and on further heating turns into viscous liquid. When the polythene is cooled, the molecular chains come close together again and attract one another more. The material solidifies and becomes stronger and stiffer.

With polymers like polythene made up of long-chain molecules, the process of softening on heating and hardening on cooling can be repeated almost indefinitely (provided that the polymer is not heated so strongly that it decomposes). Plastics which have this property are known as *thermoplastics*. Other thermoplastics are PVC and nylon.

Some plastics behave quite differently from the thermoplastics, when they are heated. A good example is Bakelite. This is really a trade name for a type of plastic made by the chemical condensation of phenol (C_6H_5OH) and formaldehyde (HCHO). It was by causing these two chemicals to react together that the first fully synthetic plastics were made over fifty years ago by the Belgian chemist Leo Baekeland. The word Bakelite is derived from his name. Similar products with different trade names are now made by many firms in the plastics industry.

When a large number of molecules of phenol and formaldehyde are heated together under the right conditions, with a suitable catalyst, they undergo a condensation reaction to form polymeric chains similar to thermoplastics.

The product is separated as a brownish powder which, when mixed with other ingredients, is known as PF moulding powder, PF being the initial letters of phenol formaldehyde. (Several plastics are commonly known by their initial letters.) PF powder is used to make shaped articles by putting it into a moulding press and subjecting it to further heat and pressure for a short time; the powder is changed to a hard glossy solid which takes up the shape of the mould. Unlike thermoplastic materials, the PF plastic cannot be softened again by heating, but remains a hard infusible solid. Plastics like this, which can be softened only once during moulding, are said to be *thermosetting*.

Plastics of this kind 'set' and become hard when heated because chemical links are formed between the polymer chains at various points along their length. This process, *cross-linking*, firmly binds the chains to one another throughout the mass of the material. So, instead of an assembly of separate long chains, the polymer consists of a three-dimensional network in which each original chain molecule is chemically linked with all the others. The whole mass of the polymer is, in effect, one enormous molecule.

The plastics industry now produces many other kinds of thermo-setting plastic. Some of the more common ones are made, like PF, by condensing formaldehyde with another chemical. Two plastics of this type are urea-formaldehyde and melamine-formaldehyde. They have the advantage over the phenol-formaldehyde materials of being light in colour. They can therefore be pigmented to give decorative pastel shades for such things as tableware, electric fittings, and radios.

How many properties of polymers can you think of which distinguish them from other materials such as metals, pottery, and glass? You can compare the structure of polymers with the structures of metals and silicate minerals by reading chapter 9 in the *Handbook for pupils*.

THE PLASTICS INDUSTRY

The plastics industry is composed of many different kinds of company. The companies that make the polymers — the plastics raw material — are essentially part of the chemical industry. There are also a large number of fabricators or converters who change the raw materials into finished products by moulding, extruding, and various other processes. The makers of the special plant and equipment used in these processes are also regarded as part of the plastics industry. The industry as a whole is, therefore, very complex and overlaps the chemical and engineering industries.

Carrying out polymerization reactions on an industrial scale, the job of the manufacturers of plastics raw materials, is much more complicated than making small amounts of polymer in the laboratory. Since the properties of plastics depend on their chemical structure, plastics materials must be made of a constant and reproducible composition.

Industrial polymerization processes require a great variety of plant and equipment. For example, to make polythene by the original process you need a plant able to withstand pressures of 1000 atmospheres or more. In a

PVC plant, the vinyl chloride monomer is polymerized in pressure vessels, called autoclaves, in which the volatile monomer, mixed with water, has to be stirred continually during the reaction. To make clear acrylic sheet, the liquid monomer, containing a small amount of catalyst, is run into glass moulds at ordinary pressure and polymerized by heating the moulds in large ovens under carefully controlled conditions. These few examples of the kind of plant used in making plastics show that this section of the industry involves not only chemical processing on a large scale but also much engineering and technological skill.

The products of the plastics raw materials manufacturers are sent to the fabricators in various forms. (Plate 13.) They may be powders, small chips or granules, and occasionally viscous liquids. Some plastics materials are made in such semi-finished forms as sheet, film, rod, or tubing.

The usual way of making shaped, finished articles from plastics is in some form of heated mould under pressure. The methods used depend on whether the material is thermoplastic or thermosetting. The commonest types of moulding process are injection moulding (figure 95), blow moulding (figure 96), vacuum moulding (figure 97), extrusion (figure 98), and compression (figure 99). Plastics can also be extruded through shaped holes to produce such equipment as piping. The machinery required may cost thousands of pounds, and the plastics materials — compared, for example, with wood — are expensive. But so efficient are the machines at turning out precisely shaped articles at a very quick rate that articles made from plastics are usually cheaper to buy than similar articles in other materials (figure 100).

Look at a variety of plastic objects in your home. Can you decide which process was used to mould them by studying their shape and any seams or surface marks?

Which of the methods of making shaped articles from plastics would you use to pro-

Figure 95
Injection moulding — granulated thermoplastic is fed through a hopper into a heated cylinder. The cylinder has a plunger at one end and a hole at the other. The softened thermoplastic is forced by the plunger through the hole into the unheated mould. The thermoplastic cools, the mould opens, and the finished article is ejected. The photograph shows a test injection moulding in high density polyethylene being examined.
BP Chemicals Ltd

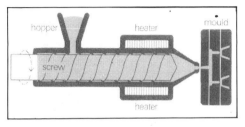

Figure 96
Blow-moulding — this process is used to produce bottles and other hollow containers with a small opening. A length of softened plastics tube is inserted in the hollow cavity of the vertically split mould. As the mould closes round the tube, it seals off one end. Compressed air, forced in at the other end, blows the tube into shape. The mould opens and the hollow container is ejected, in this illustration a milk can.
BASF

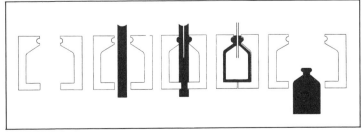

Polymers from petroleum

Figure 97
Vacuum forming — heated plastics sheet is drawn over a mould, a vacuum is created in the space between the sheet and the mould, and the plastics sheet is sucked onto the mould shape. Here a box is being removed from the machine. Pressure forming is the reverse process.
BP Chemicals Ltd

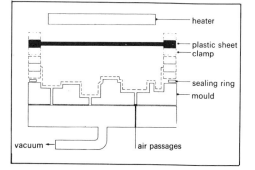
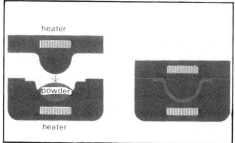

Figure 98
Extrusion — granulated thermoplastic is fed into a heated cylinder and is continuously moved forward by a revolving screw. Upon reaching the end of the screw the thermoplastic is forced through a hole whose shape determines the resultant shape of the material. The photograph shows the adjustment of the extruder head in the production of a gas distribution main, from polyethylene.
BP Chemicals Ltd

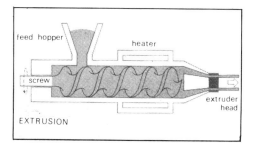

Figure 99
Thermosetting plastics. Compression moulding — a two-piece mould is heated with steam and the lower half is filled with a pellet of the thermosetting moulding powder. The pellet softens, the upper half of the mould is compressed onto the lower, and the powder is forced into the mould shape. Cross-linking occurs and the mould opens to eject the finished article. Here electrical equipment made of Bakelite is being lifted from a compression mould.
Bakelite U.K. Ltd

Figure 100
The lamination process can be used for both thermoplastic and thermosetting plastics. The plastics material can be impregnated onto paper, or it can be used to bond layers of glass fibre, chipboard, or cloth. Here a pressed pack of laminate sheets is being received from the unloading frame.
Bakelite U.K. Ltd

duce an ashtray, a drainpipe, a wash-basin, a tray, a light switch, or a sack for fertilizer?

Spinning polymers to form fibres

In the Garden of Eden, Adam and Eve clothed themselves with fig leaves. Early man, living in harsh climates, wore animal skins. Then, so it is thought, people began collecting wool fibres, torn from the fleece of sheep by bushes and scrub. At first they probably pressed the fibres together to make felt. Certainly nomads in the steppes of Russia and Central Asia have been making felts for thousands of years, and using them for tents and clothing. Then people learnt to spin the wool into a continuous yarn. In China they learnt to unravel the cocoon of the silkworm into a fine silk thread many hundreds of metres long. In India they discovered how to make another kind of thread from the seed hairs of the cotton plant, and also sacking from jute. In Egypt they made linen from

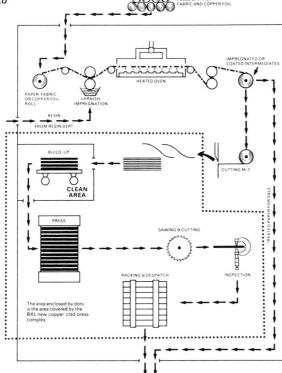

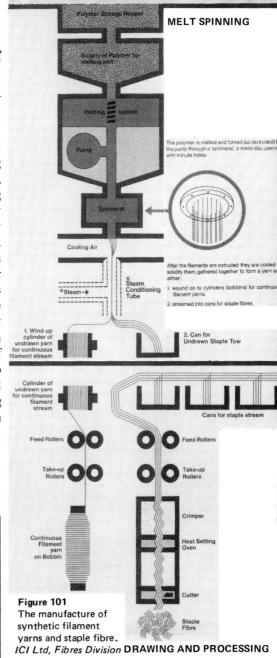

Figure 101
The manufacture of synthetic filament yarns and staple fibre.
ICI Ltd, Fibres Division

flax. Man relied on natural fibres, and looked for fibres that were strong, supple, and comfortable to wear.

The idea that man might make fibres artificially, instead of gathering them ready-made from nature, came from observing the silkworm (figure 101). To protect itself when it pupates, the silkworm forces out a liquid which hardens into a single thread on contact with air. The silkworm winds the thread round its body and is thus enclosed in a cocoon.

Polymers from petroleum

The manufacturers of synthetic fibres use what is called a spinneret which, in the way it works, imitates the spinneret in the silkworm and spider. The spinneret consists of a metal plate pricked with tiny holes. A molten substance forced (or extruded) through the holes emerges in the form of filaments (figure 102).

Imitating the silkworm, scientists find it useful to stretch the filaments after they are extruded (figure 103). The stretched filaments are both thinner and stronger. The increased strength of the filaments is caused by the re-alignment of the long-chain molecules to give a more crystalline structure. A variety of yarns can be made by cutting the continuous filament into short pieces from 2 to 20 centimetres long (called staple) and then twisting them together as in the traditional spinning of wool and cotton. From the staple, it is possible to produce yarns with a rugged texture that can be used in materials for suits, or yarns as fine as high-grade cotton.

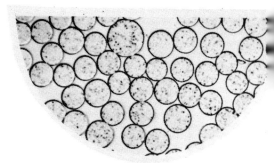

Figure 103
Cross-section of nylon under the microscope, x 500.
Courtaulds Ltd

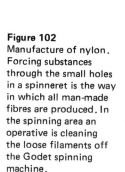

Figure 102
Manufacture of nylon. Forcing substances through the small holes in a spinneret is the way in which all man-made fibres are produced. In the spinning area an operative is cleaning the loose filaments off the Godet spinning machine.
ICI Ltd, Fibres Division

The chemistry of synthetic fibres has much in common with that of plastics. Indeed, most of the newer polymers used to make fibres also have uses as plastics. Nylon, for example, is used to make gear-wheels and other machine components, and Terylene is used to make adhesive tape and photographic film. However, not all plastics are suitable for making fibres. For conversion into a textile fibre, there are certain properties which a polymer should possess.

The polymer molecules should be long and straight with no branching of the chains. The repeating units in the chains should be arranged symmetrically and there should be no bulky side groups. There should also be evenly spaced groups of atoms which give strong intermolecular forces which help to form a highly crystalline structure when the fibres are extruded and stretched.

Man-made fibres now account for a substantial part of textile production, and their use has affected a number of industries. Obviously this effect has been greatest in the clothing industry. Often textile machinery has had to be redesigned; spinning, knitting, or weaving techniques have been modified in order to make the fabrics. Here, a great advantage of staple fibre is that, whether it be viscose, nylon, or any other man-made fibre, it can be spun on all types of textile machinery — either alone or blended with natural fibres (figure 104). Yarn spun on cotton-spinning machinery tends to resemble cotton; yarn spun on wool-spinning machinery tends to resemble wool; and yarn spun on flax-spinning machinery tends to resemble linen. Costs also come into it. When they are intended for cutting into staple, the filaments can be extruded in much greater quantity.

Because nearly all the man-made fibres are strong, crease-resistant, and not attacked by moth, they have presented a challenge to manufacturers who use natural fibres. Consequently, great efforts have been made to improve the properties of natural fibres. Wool, for example, is often made moth-proof, and techniques have been devised for putting permanent creases into woollen fabrics. Other industries that have been affected by man-made fibres are dyeing and dry-cleaning. Fabrics made from synthetic fibres are relatively non-absorbent and therefore have to be dyed and finished by new and complex processes. To clean them new solvents have had to be found. The housewife, too, has to be careful about washing and ironing because fabrics can be damaged by too high a temperature (figures 105, 106).

Which do you prefer to wear: a terylene or a cotton shirt, a woollen sweater or one knitted in nylon? Why do you think that many clothes are made with mixtures of natural and synthetic fibres?

Figure 105
In addition to fabrics, man-made fibres have many industrial uses: light, strong nylon safety nets used during the construction of the Runcorn–Widnes bridge over the river Mersey. The nets weigh much less than comparable sisal nets.
ICI Ltd, Fibres Division

Figure 106
'Terram', 75 per cent polypropylene / 25 per cent nylon, being unrolled over roughly levelled ground to form the base for temporary roadways.
ICI Ltd, Fibres Division

Figure 104
Staple fibre — that is, raw wool or cotton or short lengths of man-made fibre — is spun into continous yarns on machines such as this.

Polymers from petroleum

Chapter 7
FERTILIZERS

There is an urgent need to grow more food for the rapidly increasing population of the world. One person in every eight does not have enough to eat, and many more do not get the right kind of food. One person in every eight adds up to a lot of people, and, if this were not enough, the population of the world is increasing by about 60 million every year. Over 200 years ago Jonathan Swift wrote in *Gulliver's Travels:* 'Whoever could make two Ears of Corn, or two Blades of Grass to grow upon a spot of ground where only one grew before; would deserve better of Mankind and do more essential Service to his Country, than the whole Race of Politicians put together.' Swift was bitter about politicians, but the main point is clear, and the problem is more pressing than ever before. People look to science for means of solving it.

One thing that is being done is to bring more land under cultivation. Most of the world's uncultivated land consists of waterless desert, swamp, mountain, and impenetrable jungle, and the task of cultivating it is formidable. Moreover, thousands of acres of cultivable land are lost each year because of soil erosion, desertification, industrial development, and the building of towns and roads. It is hard to make much headway. There is, however, one immediate target – to increase production from the land that is already being used for growing food. Three ways of doing this are: planting seeds that give a greater yield of crops; feeding the soil with enough fertilizer for the crops to grow abundantly; and protecting the crops, with chemicals, against weeds, insect pests, and plant diseases.

Already these ways of increasing food production are being followed on a large scale. Agricultural research stations are breeding seeds that double or treble the yield of crops, and the manufacture of fertilizers and farm chemicals has grown into a thriving industry. The success of these remedies depends on the co-operation of the chemist, the biologist, and the farmer.

You can read more about the world food problem and the problems of malnutrition in Chapter 11 of the *Handbook for pupils*. Chemists have played an important role in the development of pesticides, herbicides, and fungicides to protect crops, and they have also helped to discover food additives to preserve food and improve its taste and appearance. These aspects of the chemist's work are also described in Chapter 11 of the *Handbook*.

FOOD FOR PLANTS

The plant is a natural chemical laboratory. Using light from the sun as a source of energy, the plant combines carbon dioxide from the air with water from the soil to make sugar and oxygen. The sugars so formed provide the plant with a source of energy. Plants are also able to take up through their roots compounds such as nitrates and sulphates which are dissolved in the water in the soil. These are converted into amino-acids and eventually form the proteins of which the body tissues of both plants and animals are composed. Man, in common with other animals, depends on plants for his food – either directly through eating plants or indirectly through eating other animals.

Eventually, it is true, we may be able to make some of the organic chemicals which we need for food directly from inorganic chemicals. As yet, however, this is only possible on a very small scale, and in the laboratory.

But there is a very great difference between plants growing wild in nature and plants growing on cultivated ground. In nature, the dead plants return their substance to the soil. On cultivated ground, they do not. Instead, they are harvested to produce food, and a second crop is planted in their place. But the inorganic salts in the soil on which the plants feed are not inexhaustible. If nothing is returned to the soil, it will not go on supporting fresh crops of plants year after year. There is an old saying among farmers that 'what is taken out of the soil must be put back'.

What it is that plants take from the soil, and what therefore has to be put back, has only been known for about the last hundred years. The starting-point for much of the more recent research was the work of the German chemist Baron Justus August von Liebig (1803–73), who analysed the ash left from the burning of plants — an investigation you can easily repeat. As a result of his work and the work of other scientists, it is known that the chief needs of plants from the soil are various compounds containing nitrogen, phosphorus, potassium, and calcium. Most plants also require much smaller quantities of compounds containing sulphur, magnesium, iron, chlorine, zinc, manganese, boron, copper, and molybdenum (figure 107).

In normal soils, there is a large reserve of all the plant foods listed above. But the plant roots can absorb these foods only if they are dissolved in water, and the supply of soluble salts in soils is usually insufficient for the needs of a crop. In Britain it is usually necessary to add only those compounds containing nitrogen, phosphorus, potassium, and calcium. Elsewhere, however, such additional elements as boron, magnesium (a constituent of chlorophyll), and manganese may also have to be added (plate 14).

The plant foods which the farmer adds to the soil are called fertilizers. Basically there are two kinds of fertilizer — the organic fertilizers and the inorganic salts (figure 108). The bulk of organic fertilizer comes in the form of farmyard manure, consisting of the dung and urine of farm animals and the litter (usually straw) on which they are bedded. The manure, when put into the land, performs two important functions. First, it acts as a soil conditioner, improving the drainage and aeration of heavy soils and the water-holding capacity of sandy soil. Second, it provides plants with all the food they need, including a variety of trace elements.

Although manure is an excellent fertilizer, there is not enough of it to satisfy the needs of most crops. Manure returns to the soil only a small part of the plant foods leached away by drainage waters, or removed in the meat, milk, wool, vegetables, and cereals which are sold off the farm.

Figure 107
Bean plants growing in various nutrient solutions.
Rothamsted Experimental Station

Figure 108
Since Liebig's analysis of plant ash over a hundred years ago, much has been learnt, using new scientific techniques, about the food requirements of plants. Here at Rothamsted plants are being examined with a Geiger counter to find out their uptake of a specially prepared radioactive phosphorus compound.
Central Office of Information

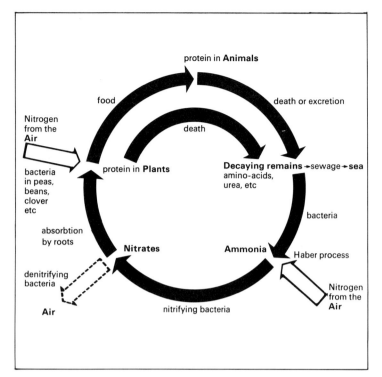

Figure 109
The Nitrogen Cycle: although, when a plant absorbs nitrates from the soil, the nitrates undergo a number of complex chemical conversions, most of them return to where they started — as nitrates in the soil.

Some of the nitrogen which a crop needs comes from the decaying remains of previous crops. Some of the nitrogen comes from the activity of certain plants, called the leguminous plants (clover, peas, and beans, for example), whose roots contain special bacteria able to convert the nitrogen in the air to chemicals which can be used for plant growth. Finally, if the vegetable remains and nitrogen fixed from the air are not sufficient to support a crop, nitrogenous fertilizers must be added.

The need to add to the natural reserves of nitrogen in the soil has been recognized for a long time (figure 110). A century ago, there was a flourishing fertilizer trade based on a natural fertilizer — the sea-bird droppings which had accumulated over centuries on the rocky coasts and islands of South America. Eventually, however, supplies of *guano*, as this manure was called, became exhausted. Towards the end of the nineteenth century, farmers turned to the sodium nitrate deposits of Chile to provide them with nitrogenous fertilizer; but again the supply was limited.

Meanwhile the chemical industry was demanding more nitrogen to make dyestuffs and high explosives such as dynamite. Agriculture and industry were competing with one another for the available nitrogenous compounds. This, then, was the nitrogen problem: unless new sources of nitrogen could be found, the world would be faced with starvation.

Thus plant food, in the form of inorganic salts, must be put back into the land to balance the loss. These salts are sometimes called 'artificial' fertilizers, but there is nothing artificial about the elements nitrogen, phosphorus, and potassium which they contain. They are the same plant foods, having much the same effect on a crop, as those provided in farmyard manure. Huge tonnages are required throughout the world every year, and it is here, in the large-scale preparation of inorganic fertilizers, that the chemist and chemical engineer make a vital contribution to agriculture (plate 15).

Why do we hear and read more about the nitrogen problem than about the phosphorus, potassium, or calcium problems?

THE NITROGEN PROBLEM
More than any other plant food, compounds containing nitrogen influence the yield of a crop, and especially such leafy crops as grass, cereals, and cabbage (figure 109).

An obvious source of nitrogen was the atmosphere. It had been known for at least a hundred years that nitrogen made up about four-fifths of the gases of the atmosphere. If this nitrogen could be extracted from the atmosphere and fixed into compounds that could be used as fertilizers, the threat of a nitrogen shortage would be removed for ever. To take oxygen from the air and fix it into compounds is easy; it happens whenever something burns. But nitrogen is very much less reactive. Chemists had to work hard to find an answer.

Two processes for fixing nitrogen were developed around 1900. Both made use of the tremendous heat produced by an electric arc. In the Cyanamide process, pure nitrogen was passed over strongly heated calcium carbide to give calcium cyanamide. In the Birkeland and Eyde process, nitrogen and oxygen were combined as nitrogen oxide in an electric arc. Both processes were expensive because they needed a great deal of electrical energy, and in 1909 another, cheaper method of fixing nitrogen was discovered.

Fritz Haber's solution of the nitrogen problem

The most important solution of the nitrogen problem was discovered by Fritz Haber. He was the son of a merchant of Breslau (now Wroclaw), a town in what was then German Silesia and is now Poland. After studying chemistry, the young Haber went into business but did not like it much. He gave it up and obtained a post as lecturer at the technical college of Karlsruhe (figure 111). There he specialized in electro-chemistry and physical chemistry. He knew about all the efforts then being made to produce fixed nitrogen, and in 1904 he too began to work at this problem. He became interested in the possibility of combining nitrogen and hydrogen to form ammonia. The equation for this reaction is:

$$N_2(g) + 3H_2(g) \rightarrow 2NH_3(g)$$
$$\Delta H_{932K} = -109 \text{ kJ}$$

In practice, an equilibrium is set up:

$$N_2(g) + 3H_2(g) \rightleftharpoons 2NH_3(g)$$

At ordinary pressures, hardly any of the nitrogen and hydrogen combine. The equilibrium is over towards the left ($\leftarrow$). However, since the volume of ammonia is only about half the combined volumes of nitrogen and hydrogen, increasing the pressure (and thereby decreasing the volume) causes the equilibrium to shift to the right ($\rightarrow$) and so yields a greater quantity of ammonia.

Temperature, too, affects the equilibrium. Since heat is given out in the reaction, the lower the temperature, the greater the yield of ammonia. However, the lower the temperature, the longer the time needed for the reaction to reach equilibrium, and in a manufacturing process speed is important. Therefore a compromise is necessary between the lowness of the temperature and the speed at which equilibrium is reached.

At first Haber's work was theoretical: calculating the temperatures and pressures necessary to produce the greatest quantity of ammonia. He thought that, to produce even a small amount of ammonia, there would need to be a pressure of several hundred atmospheres and a temperature of several hundred degrees Centigrade. Walter Nernst disagreed with Haber, as he had been able to obtain a very small amount of ammonia by reacting nitrogen and hydrogen at fairly low pressures. Haber revised his calculations. He also began to look for a catalyst to make the reaction faster: osmium gave the best results.

Figure 110
The effect of fertilizer on a potato crop — only the plant in the middle has been fed with fertilizer.
Fisons Studio, Fertilizer Division

Figure 111
Fritz Haber (1868–1934) at work in his laboratory at Karlsruhe.
Dr. L.F. Haber

Fertilizers

In 1908, four years after he had started on this work, Haber found that, at a temperature of 500–600 °C and a pressure of 175 atmospheres, nitrogen and hydrogen combined in the presence of a suitable catalyst to give a yield of 8 per cent by volume of ammonia. He was fortunate that the pressure needed to be no higher, because the laboratory compressor at Karlsruhe had an upper limit of 200 atmospheres. He was also fortunate to have able and skilful assistants, because the design of the apparatus was complicated, and Haber himself was a butterfingers.

During the first half of 1909, several demonstration runs were made. These not only proved the value of Haber's painstaking theoretical work, but also showed that the process was a practical one and that it could have a revolutionary effect on industrial chemistry. This process finally displaced all the other methods of nitrogen fixation.

When building his laboratory model, Haber made use of two ideas (figures 112 and 113). The first of these was *continuous flow*. The reactants are fed in continuously at one part of the apparatus and the products are continuously drawn off at another, just like the conveyor-belt system. In Haber's model, the mixture of hydrogen and nitrogen was continuously pumped through the reaction chamber or converter, with the result that the unreacted gases from the converter were always replenished by a fresh supply. This made up for the ammonia which had been withdrawn. The second idea that Haber made use of was that of *heat exchanger*. An exchanger was built into the apparatus and the ammonia was made to give up its heat to the incoming gases. This saved valuable energy. Continuous gas flow and heat exchange proved to be of fundamental importance. They are still used, along with the principle of catalytic action at high pressure and temperature, in all the ammonia plants that have been built. They have also been used in many other chemical processes.

Haber was awarded the Nobel Prize for chemistry in 1918 for his discovery of the ammonia synthesis.

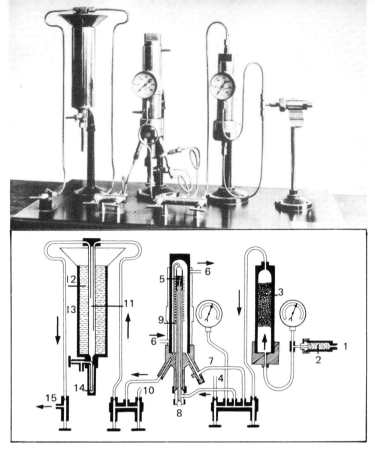

Figure 112
Haber's experimental apparatus for the synthesis of ammonia.
Deutsches Museum, Munich

Figure 113
Diagram of Haber's experimental apparatus. Key to Diagram:

1. entry of compressed gases from the circulating pump (nitrogen and hydrogen with traces of oxygen)
2. heated platinum asbestos to catalyse the reaction between hydrogen and traces of oxygen and so produce water
3. drier to remove the water produced in the reaction above
4. feed-in for fresh supply of nitrogen and hydrogen
5. converter
6. water cooler
7. pressurizer
8. heating elements
9. electrical heating
10. tube for the ammonia-holding gases
11. flow tube
12. cooling bath
13. heat exchanger
14. offtake for liquid ammonia
15. return of the high pressure gases to the circulating pump

---

Can you think of examples of equipment you have seen or used at home or in the laboratory which work on a continuous flow principle or involve heat exchange?

Synthesis of ammonia on an industrial scale

In 1909 a large German firm manufacturing dyestuffs, Badische Anilin und Soda-Fabrik, bought the rights to the Haber process. So began the formidable task of developing the laboratory apparatus at Karlsruhe into a large plant. It was brilliantly accomplished by Carl Bosch. Bosch's father owned a plumber's business, and this background was most useful to his son, who had worked as a fitter for some time before going to the university

Figure 114
A painting of the first synthetic ammonia plant at Oppau in 1914. On the left, by the river, is the coke plant. The coke was reacted with steam to produce hydrogen and the hydrogen combined with nitrogen from the air to make ammonia. On the right, at the back, is the gas-holder for ammonia.
BASF

to read chemistry. In 1899 he joined Badische Anilin. He quickly made a reputation as an outstanding organizer and chemical engineer.

Bosch and his team had to deal with three main problems in building a plant to work the Haber process. The first was to design a converter to withstand the high temperatures and pressures. Such severe operating conditions had not been met before, and many metallurgical problems had to be solved. In 1913 Bosch designed a steel converter which weighed 3½ tonnes and was 7 metres high. Nowadays a converter weighs around 170 tonnes and is about 20 metres high.

The second problem was the choice of catalyst. Those used by Haber were satisfactory, but expensive. Bosch had this problem thoroughly investigated, and by the end of 1912 some 6500 experiments had been carried out. His staff eventually found that an activated iron catalyst containing oxides of potassium, aluminium, and calcium gave the best results; this type of catalyst is still used and has proved itself satisfactory at a working pressure of 300 atmospheres.

Finally, Bosch had to devise cheap ways of getting hydrogen and nitrogen. At first, hydrogen came from the electrolysis of water, and nitrogen from the liquefaction of air. Both these processes were too expensive for large-scale operations. To make hydrogen, Bosch invented a process whereby steam is blown over coke and the resulting water gas is catalytically separated into hydrogen and carbon dioxde. A few years later, his process was supplemented by another in which air is blown over coke, and the nitrogen-rich gas is freed from carbon monoxide before it is pumped to the converter.

The final answer to these and other problems took many years to find. However, in 1912 it was decided to build an ammonia plant, and towards the end of the next year a complete works was operating at Oppau on the Rhine (figure 114). It had a capacity of 30 tonnes of ammonia a day and could make 30 000 tonnes of sulphate of ammonia a year. With the success of this plant — a forerunner of many others — it became evident that the menace of famine caused by lack of nitrogenous fertilizers had finally been overcome. Bosch, through building the plant, had created a new branch of the chemical industry based on high-pressure techniques. He began to apply these new techniques to related industrial processes. For his achievements in high-pressure chemistry, Bosch shared the Nobel Prize award in 1931.

Starting with Haber's calculations and ending with the manufacturing plant at Oppau, this transformation of a theoretical idea into a practical reality is a notable example of the great practical benefits gained when the skills

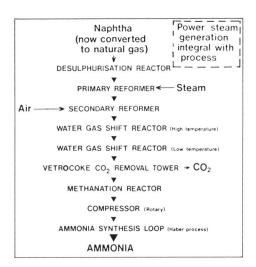

Figure 115
Stages in the synthesis of ammonia.
ICI Ltd, Agricultural Division

Figure 116
The three low pressure ammonia units at Billingham.
ICI Ltd, Agricultural Division

of the chemist and the engineer are combined. Thanks to the success of Haber and Bosch, the nitrogen problem was solved and the threat of world starvation was averted.

Today nearly all nitrogen compounds are made from ammonia which has been synthesized by the Haber process. In modern plants, synthesis gas (the mixture of nitrogen and hydrogen needed for the process) is no longer made by blowing steam over coke, but by newer processes. For example, in 1954 a method was introduced based on the reaction of naphtha (a petroleum product – a mixture of C_6 and C_7 liquid hydrocarbons) with steam and air to provide hydrogen and nitrogen in the correct proportions. But now in Britain synthesis gas is made from natural gas from the North Sea (figure 115).

In 1975/6, 43.3 million tonnes of combined nitrogen were applied to the soil in fertilizers, but even this huge tonnage is not enough. Africa, Asia, and Latin America are using very little nitrogen at present, yet they must grow more food because their population is increasing rapidly. The world will have to produce even more fixed nitrogen than it does today (figure 116).

The success of the ammonia synthesis process depended on the work of Haber who was more concerned with the chemistry of the synthesis, and on the work of Bosch who dealt with the engineering of the plant. If you were to make a career of chemistry would you prefer to work in a research laboratory or to be a chemical engineer?

NITROGEN FERTILIZERS

In some parts of the world, especially in the U.S.A., ammonia is liquefied and injected into the soil directly, but it is usual in the manufacture of fertilizers to convert ammonia into different compounds. Some ammonia is reacted with sulphuric acid to produce fine white crystals of ammonium sulphate. Because this compound makes the soil acid, an equivalent amount of chalk must be added when ammonium sulphate is used as a fertilizer. In the soil most of the ammonium ions are converted by soil bacteria into nitrate ions before they are absorbed by the plants.

In Britain the most popular nitrogenous fertilizer is ammonium nitrate. To make this salt some of the ammonia is converted to nitric acid which is then neutralized with more ammonia to make ammonium nitrate. As with ammonium sulphate, most of the ammonia in ammonium nitrate is converted into nitrate by soil bacteria, but because this conversion takes time, the action of the nitrogen is spread over a longer period. If the nitrate is not absorbed rapidly by the plants, it is soon washed out of the soil, and for this reason nitrate fertilizers applied in late autumn or winter to uncropped land are largely wasted. They are therefore usually applied shortly before sowing or planting a crop in the spring, or to a growing crop in the form of what is called a top-dressing (because it is not ploughed in) (figure 117).

CALCIUM FERTILIZERS

Calcium is applied to the soil almost entirely as calcium carbonate, usually in the form of ground chalk or limestone. The popular name for this fertilizer is 'lime'. Strictly speaking, lime is not a fertilizer but a soil conditioner. Although calcium in small amounts is an essential plant food, the principal function of the calcium carbonate is to reduce the acidity of the soil. Most plants will thrive best when the pH value of the soil is about 6.5, even though some crops — for example, oats, rye, and potatoes — will grow well in a soil with a higher acid content. (The presence of such weeds as sheep's sorrel and corn spurry shows that the soil is acid.) The pH value of the soil also influences the functioning of soil bacteria, which play an important part in converting non-usable substances into others which can be used by plants.

PHOSPHATIC FERTILIZERS

Phosphates are important because they stimulate root development and the early growth of crops. For this reason they are added to the soil at sowing time. Their use has to be timed carefully, however, because they can form insoluble compounds which cannot be absorbed by plants. Phosphates are also used to bring forward the ripening of crops and to stimulate nitrogen-fixing bacteria. They are particularly important on heavy, wet soils and for encouraging clover growth in pastures, and they are traditionally used in large quantities in growing turnips.

At one time, animal bone provided the main source of phosphorus, but now most of it comes from a rock containing phosphate which is found chiefly in the United States, North Africa, and Russia. The mineral can

Figure 117
A modern type of storage silo. The product is brought on a conveyor band incorporated into a swinging arm. It is removed by a mechanical grab and put on conveyor bands which transport it to the packing sheds.
ICI Ltd, Agricultural Division

be ground up and used as a fertilizer on its own, but in this form it is effective only in acid soils and in areas of high rainfall. Accordingly most of it is treated with sulphuric acid to make a more soluble product called superphosphate. Phosphatic fertilizers also include basic slag, a by-product of the steel industry, and ammonium phosphate, which is prepared by treating ammonia with phosphoric acid, and which supplies nitrogen as well as phosphorus to crops.

POTASSIUM FERTILIZERS

Potassium is necessary if plants are to make efficient use of all the available nitrogen in the soil. It also helps plants to make sugar and starch, and to stand up to drought and disease. Potassium generally improves the quality of crops. Soluble potassium compounds are fairly abundant in clay soils, but are usually deficient in chalk or sandy soils.

Potassium fertilizers are generally referred to as 'potash' — the common name for potassium compounds. A widely used potassium fertilizer is potassium chloride, sometimes known among farmers by its old chemical name of muriate of potash, which is produced from natural deposits of potassium salts. Large deposits of sylvinite (a mixture of potassium and sodium chlorides) have been discovered in north-east Yorkshire and techniques have recently been developed for mining the rock and separating the potassium chloride for use in fertilizers. A purer but more expensive potassium fertilizer is potassium sulphate, prepared by treating the chloride with sulphuric acid. Farmers use this for potatoes and crops grown under glass.

APPLYING FERTILIZERS TO THE SOIL

Usually a crop requires more than one plant food at a time, so obviously the farmer can save time, labour, and machinery if the foods are mixed together and he can apply them all at once.

The mixing is done by machines in the fertilizer factory. The required ingredients are weighed, ground, and put together in the correct proportions in special rotary mixers.

A further improvement has been the introduction of granular fertilizers. These are easier to use and store than powder or crystalline materials. The fertilizer comes out of the mixer in the form of a hot sludge, and this is granulated, cooled, and dried by rotation in very large drums. The finished granules are hard, and roughly spherical.

In Britain, most nitrogen, phosphorus, and potassium (NPK) fertilizers are now used as mixtures — although they are commonly known as 'compound' fertilizers. Nitrogen fertilizers on their own ('straight' fertilizers) are used principally during spring and summer for top-dressing cereals and cabbage crops already growing, and for putting on grassland at intervals during the growing season. The only other fertilizer which is usually applied 'straight' is basic slag, and this all goes on grassland (figure 119).

To summarize the information in this chapter, you might draw up a table or flow diagram showing the various sources of the nitrogen, phosphorus, calcium, and potassium required for fertilizers; and then indicate how these are combined to form fertilizers together with hydrogen, oxygen, and sulphur. You may need to consult other reference books to obtain all the information.

Figure 119
Spreading granular fertilizer.
ICI, Agricultural Division

Figure 118
Granules of 'Nitro-top' fertilizer, showing the consistency of size. This regularity helps in spreading the fertilizer evenly. Note how small the granules are.
Fisons Studio, Fertilizer Division

FERTILIZERS AS POLLUTANTS

There are dangers in the regular large-scale use of inorganic fertilizers. They may change the soil and reduce its fertility. One of the results of these changes is that the soluble salts in the fertilizer are not held in the soil, but are washed away into streams and rivers. Nitrates in rivers may then enter domestic water supplies and harm very young children, by making the blood less able to carry oxygen. So far this kind of pollution has not become a serious problem in Britain. In any case the presence of nitrates in rivers does not depend only on the use of artificial fertilizers. It has been reported in some areas that as much nitrate is washed from unfertilized land as from fertilized land; some rivers show little or no increase in nitrate concentrations despite large increases in the amount of fertilizers applied.

Another hazard can arise when soluble, inorganic fertilizers are washed from the soil and carried by rivers into lakes; this is that the lakes become 'eutrophic', meaning that they become rich in nutrients which stimulate the growth of plants. If the problem becomes acute the lake is choked with growths which then die and rot, using up the dissolved oxygen in the water. This kills the fish and other animals and converts the lake to a lifeless swamp. Again, this is not yet a serious problem in Britain, but a well-known example of a lake affected by eutrophication is Lake Erie. However, it is not only fertilizers which cause this process: the phosphates naturally present in sewage effluents, supplemented by the phosphates added to washing powder, make a major contribution to the raised nutrient levels in rivers and lakes.

Chapter 8
CERAMICS AND GLASSES

CERAMICS AND GLASS FROM CLAY AND SAND

The plastics and fibres described in Chapter 6 are new materials which have only been common in our homes in the last twenty-five years. Ceramics and glasses are traditional materials which have been made by man for thousands of years.

The name 'ceramic' comes from a Greek word meaning pottery, or burnt stuff, and potters have fashioned and fired clay to make bricks, vessels, and figures since prehistoric times. Most ceramics are made from clay, which can be dug from the ground, purified, moulded when wet, dried, and then heated in a fire to harden (figure 120). Most glass is produced from sand (silicon dioxide), limestone (calcium carbonate), and soda ash (sodium carbonate made from salt) by melting these substances together in a furnace. The liquid glass is cooled until it is thick enough to mould, then shaped and cooled further until it sets solid (figures 121 and 122). Notice the skill needed for this craft.

Thus everyday ceramic and glass objects are made from common materials available cheaply in large quantities. Both ceramics and glasses need, during manufacture, to be fired at a high temperature in a furnace, and chemically they have a similar composition: giant structures of silicon and oxygen atoms.

Why were the techniques for making pottery and glass developed thousands of years before methods of making synthetic plastics and fibres were discovered? How much theoret-

Figure 120
China clay face during excavation in Cornwall.
Nicholas Horne Ltd

Figure 121
Traditional glass blowing. A lead crystal glass is being blown by mouth.
Glass Manufacturers' Federation

Figure 122
Stages in the manufacture of glass in eighteenth century France. Note the similarity of the 'chair', in which the craftsman sits, to the one in the previous figure.
From The French Encyclopaedia, 1772.
The Mansell Collection

Ceramics and glasses

ical chemistry do you need to know before making a clay pot, or a plastic beaker?

The structure of clay

Clay, and the ceramics made from it, are similar in composition to the silicate minerals such as mica and asbestos. (The structure of these minerals is described in Chapter 9 of the *Handbook for pupils*.) These are crystalline solids in which the atoms are arranged in an orderly way. Most clay is a mixture of several minerals, but many of the clays used to make pottery contain a material called kaolinite ($Al_2Si_2O_5(OH)_4$) in which the atoms are arranged in layers and the crystals form in thin plates. The plasticity of clay, which enables it to be moulded into intricate shapes, results from the way in which the kaolinite crystals stack together.

When the clay is wet, the water acts as a lubricant; this allows the platey crystals to slide over each other. When the clay is dried it becomes rigid because the crystals stick to each other. Although dry clay is rigid it is still very easily broken and it must be heated to a temperature of around 1000 °C to harden. During this firing a complicated series of changes takes place: new minerals are formed and some of the substances present in the clay combine to form glasses. The ceramic which results consists of many minute crystals of silicate minerals bonded together with glass.

The structure of glass

A lump of glass is rigid, it behaves like a solid, and yet on an atomic scale it has the appearance and structure of a liquid. Most common glasses are largely composed of silicon dioxide which is found in nature in a crystalline form as quartz. Sand is almost pure quartz which can be melted above 1700 °C. The hot molten silica is very thick and viscous, and as it is cooled back to its melting point the atoms cannot move freely enough to return to the ordered arrangement of the crystalline solid. Instead they are frozen into a disordered state (figure 123). The change from liquid to solid glass does not take place at a definite temperature; it takes place gradually as the melt stiffens on cooling. Treacle toffee is a glass, and the changes which occur when molten toffee is cooled are very similar to those observed when silicate glasses solidify.

It is difficult to handle glass at the very high melting point of silica, and so most glass is made from silica mixed with the oxides of metals. The metal oxides lower the melting point of the silica and modify its properties. Windows and bottles are made from soda-lime glass made by melting sand with the carbonates of calcium and sodium. The carbonates decompose to oxides during the melting process, so that the final glass is a combination of the oxides of silicon, sodium, and calcium (figure 124). The sparkling crystal glass used to make cut glass dishes (plate 16) and the flint glasses used for prisms and lenses are composed of silicon, lead, sodium, and potassium oxides (figure 125). The heat resistant glass used for ovenware and chemical apparatus is made from the oxides of silicon, boron, and aluminium. It is usually called borosilicate glass (figure 126).

Figure 123
Models of the atomic structure of crystalline (top) and glassy (bottom) forms of silicon dioxide, in the Glass Technology Gallery at the Science Museum, London. The basic structure of the glassy form is the same as the crystalline quartz, but the bond angles between the oxygen atoms may vary. This gives a complex random network which is typical of the structure of any glass.
Glass Manufacturers' Federation

CERAMICS AND GLASS IN THE HOME

The bricks used to build the walls of a house, the tiles which cover the roof, and the pipes which drain away waste water are all ceramics made from clay (plate 17). Inside the house, ceramic tiles may be used to decorate the walls of the kitchen and bathroom; and the sinks, basins, and toilets are often made of glazed porcelain. The roof and walls may be insulated against heat loss with glass wool, and some curtains are made with fabrics woven from glass fibres. Daylight enters through glass windows, and the glowing filament in an electric lamp is enclosed in a glass bulb. Plates, dishes, and cups are usually formed from pottery or glass, and the same materials are used to fashion many of the ornaments which decorate the home.

These uses of ceramics and glasses illustrate many of the advantages and disadvantages of this important group of materials. They are chemically inert so that the bricks, tiles, and windows are not eroded by rain water or polluted air. Likewise plates and cups are not corroded by repeated contact with hot food or washing-up water. Ceramics and glasses are hard and strong, so that plates are not scratched by metal knives and bricks can support the weight of a building.

On the other hand, they are brittle and will shatter if subjected to a sudden shock, as when a cup is dropped to the floor or a stone thrown at a window. Since most ceramics and glasses are difficult to melt, they are suitable for making ovenware; but soda-lime and lead glasses tend to crack if they are subjected to sudden changes of temperature, so glass cooking vessels are made from borosilicate glass. The transparency of glass and the fact that it is a good electrical insulator mean that it is ideally suited to making the envelopes for electric filament and fluorescent lamps.

Modern home entertainment relies on the newer ceramic materials made from compounds other than clay. In radio receivers and television sets, ceramic semi-conductors are used to make the transistors, while ferrites — ceramics derived from iron oxide — are used for their magnetic properties. Single crystals of sapphire (aluminium oxide) are used as gramophone needles, and the

Figure 124
Tumblers made from sodalime glass. Colour can be permanently fired into the glass.
Glass Manufacturers' Federation

Figure 125
Examples of lead crystal glass.
Glass Manufacturers' Federation

Figure 126
Table and ovenware made from borosilicate glass.
Glass Manufacturers' Federation

motions of the needle are converted into electrical impulses by a ceramic (lead zirconate-titanate) cartridge.

Look around your home at the ceramic and glass objects. Can you imagine a world without these materials?

CERAMICS AND GLASS IN INDUSTRY

Ceramics and glasses are used so much in industry and commerce that it is impossible to describe more than a few of their main applications. Their uses as refractories and electrical insulators are two examples.

Ceramics and glass are chemically very inert (plate 18). This is taken for granted by the laboratory chemist who can generally assume that his beakers, flasks, and basins will not interact with the reagents mixed in them. Because they are composed of oxides, ceramics and glass do not burn when heated in air. This is in contrast to metals and organic polymers, both of which combine readily with oxygen when heated, and so are usually not suitable for use at high temperatures. Ceramics are said to be *refractory* because they are very hard to melt or change in any way by heat.

Many large-scale industrial processes now involve operations carried out at high temperatures. These include the firing of ceramics, glassmaking, metal extraction, cement manufacture, and all processes which require the use of high-pressure steam. The furnaces employed for these industries have to be lined with refractory ceramic materials, because any other materials would melt or burn away.

Most refractories are made from fireclay, which in the United Kingdom is found in deposits beneath coal seams. Fireclays are more refractory than other clays because they contain a higher proportion of aluminium

Figure 127
An open-hearth steel-making furnace before use; pocket arches of high-alumina refractory.
British Steel Corporation, South Wales Group, Port Talbot.

Figure 128
A glass-making furnace after use, showing how the refractories in the 'throat' area of the furnace have been attacked by the molten glass.
British Glass Industry Research Association, Sheffield

Figure 129
Airblast switchgear and insulators in a 275 kV substation.
Central Electricity Research Laboratories

oxide combined with the silica. Bricks made from fireclay are used in the home on a small scale to line the grates of stoves and open fires. In industry they are used on a large scale to line blast furnaces, limekilns, cement and ceramic kilns, forge furnaces, and steam raising boilers (figure 127).

Although ceramics are chemically inert at low temperatures, at high temperatures they may be attacked by the hot, molten content of a furnace. Iron oxide and the slags produced in steel making, and molten glass are all highly corrosive and attack furnace linings (figure 128). Therefore in the steel and glass making industries more specialized refractories, such as bricks made of almost pure silicon oxide or of magnesium oxide, are used to build the furnace walls to give them a longer life. (Plate 8)

The transmission of electricity from power stations to homes and industry depends not only on the metallic conducting cables, but also on the electrical insulators which support them. These insulators are made of glazed porcelain or glass, because these materials are cheap and strong. They are also weather-resistant and can be fashioned into intricate shapes to ensure that in wet weather the whole surface of the insulator does not become coated with a film of moisture which would cause the insulation to break down (figure 129).

Porcelain and glass are not suitable as insulators for high-frequency alternating currents, or at high temperatures. That is why they are not used to make the insulators in spark plugs, which not only have to operate at high frequencies and temperatures, but also have to stand up to continuous vibration. Modern spark plug insulators are made from a new ceramic material, consisting largely of crystalline aluminium oxide.

Compare the refractory properties of some of the objects you have seen heated in a science laboratory, for example soda glass and pyrex test-tubes, porcelain basins, pipe-clay triangles, and silica tubing. What is the highest temperature that can be reached using a Bunsen burner?

MANUFACTURING PROCESSES
Fashioning and firing clay

The methods used to make clay ceramics depend on the plastic properties of clay when it is wet. Traditionally domestic pottery was shaped by hand or 'thrown' on a potter's wheel (figure 130) but mass-produced pottery is now moulded by machines. These use a scraper to shape the lump of clay by pressing it against a rotating plaster mould. In making plates and saucers, the plate is formed on the

Figure 130
Five stages in the throwing of a pot on a wheel:
a Placing the clay on the wheel.
b Starting the pot.
c The pot nearly finished.
d Raising the edge of the pot.
e Inscribing the pot.
Eileen Preston

a

b

c

d

e

outside of a convex mould, a process called 'jiggering'. Essentially the same procedure is used to make cups and bowls, but here the clay is forced against the inside of a convex mould, and this process is called 'jollying'.

If clay is extruded through a die, it forms a continuous column. This is very useful for making pipes and some bricks, because the column can then be cut into the required lengths. Other bricks are made by pressing the clay into steel moulds, a technique also used to make tiles (figures 131 and 132).

Hollow ware, such as teapots, ornaments, and basins, is made by a method called 'slip casting'. Slip is very runny clay which can easily be poured into moulds. The moulds are made of porous plaster which absorbs water from the clay so that a solid crust forms on the inside walls of the mould. When the crust is sufficiently thick the remaining slip is poured out, leaving a hollow object to be removed from the mould for drying and firing.

After shaping, the clay must be dried, so that no water remains, otherwise the water would turn to steam and shatter the clay during the high-temperature firing. Most commercial ware is now fired in continuous tunnel-shaped furnaces (kilns) and the clay articles are carried slowly through the kilns on trolleys running on rails. The burners which heat the kiln are placed near the middle of the tunnel. The clay ware is gradually raised to its firing

Figure 131
Making bricks by extrusion. The feeder for the extruder holds sufficient clay for one hour's production. After extrusion the column is sanded by a gravity sander to provide the characteristic surface finish.
Redland Bricks Ltd

Figure 132
The wire cutter cuts 22 bricks at one pass from the extruded clay slab, at a rate of 15 000 bricks per hour.
Redland Bricks Ltd

Figure 133
Automatically drawn tubing is used to make lighting and scientific glassware. The molten glass flows down from the tank furnace onto a slowly revolving cylindrical mandrel (a rod round which the glass is shaped). A steady flow of air is introduced through the bore and the tubing is drawn off continuously.
Glass Manufacturers' Federation

Figure 134
Diagram to show the two main processes in bottle-making. In the blow-and-blow process (A) and the press-and-blow process (B) a measured lump of molten glass drops into a blank mould, where it is blown into the shape of a hollow blank (A) or pressed by a plunger (B). In both processes the blank is transferred to a blow mould for final shaping.
United Glass Ltd

temperature as it moves towards the centre, then steadily cools as it travels on to the exit.

Apart from bricks and roof tiles, most clay ceramics are covered with a glaze. This glassy material gives the ceramic a smooth, non-porous surface which may be coloured to decorate the ware. Like other glasses, glazes are composed of the oxides of silicon and various metals. The finely powdered mixture of oxides is suspended in water, and then the ware to be glazed is either dipped into the mixture or sprayed with it. After drying, the powdery oxides are converted to a glass by a firing in a furnace. To make high-quality tableware, the clay is fired first, and then the pottery is glazed and fired again. But cheaper tableware, wall tiles, and sanitary ware are coated with glaze after the drying stage, and then the clay and glaze are fired together.

Look at different ceramic and glass objects. Can you decide which process was used to make them by studying their shape and any seams or surface marks?

Making and moulding glass

Glass articles are moulded from hot molten glass. In contrast to the fabrication of clay ceramics, the heating stages precede the shaping stages. The mixture, of oxides and carbonates, from which the glass is made is heated in a furnace, called a glass tank, lined with thick refractory walls. In the first stage of the heating the ingredients react together to make the glass. During the reaction, bubbles of gas are formed, and so, in the second stage, the glass is heated enough for it to be thin and watery, to allow the bubbles to escape. Finally the glass is cooled until it is sufficiently thick and viscous to be moulded into shape (figure 133).

Hot molten glass is formed into bottles, jars, and bulbs by a combination of blowing and moulding. A measured amount of the liquid glass is drawn into a mould, and then blown into shape by air pressure. Much table and oven glassware is fashioned by pressing the hot glass into a mould with a plunger. The materials used to make the moulds have to be carefully chosen so that they do not react with and contaminate the hot glass, and the moulds have to be made with great accuracy so that the surface of the moulded glass is smooth and clear (figure 134).

Flat glass is being used increasingly in modern architecture not only for windows, doors, and mirrors, but also for the curtain walls covering the outside of steel-frame buildings.

It is difficult, when making flat glass, to form the hot molten glass into sheets which are uniform in thickness, smooth, and shiny so that the view through them is clear and undistorted. The cheaper flat glass used for windows is made by drawing a continuous vertical sheet of glass straight from the furnace. The sheet is held at the edges by rollers, but the main part of it solidifies in contact with nothing but the air and so retains the smooth, shiny surface which it had naturally in the liquid form. Glass made in this way is never perfectly flat, and other methods have to be used to produce uniform, polished, parallel-sided sheets of glass. One method, the most modern, is called the float-glass process. The glass sheet is supported, as it cools, on a surface of hot, molten tin. The liquid glass on the bath of molten tin flows until both surfaces are smooth and level. The process is continuous: molten glass is fed from the furnace onto the tin in an endless stream, it passes across the tin surface, and as it does so it cools and solidifies. Finally it is drawn off the metal surface to be annealed.

Annealing is necessary because glass is a poor conductor of heat. When moulded glass objects are cooled, some parts cool and shrink more rapidly than others, and this sets up internal stresses which weaken the glass. To get rid of these, the article is annealed: reheated for about an hour to a temperature high enough to allow the internal stresses to even themselves out, but not so high that the glass softens and changes shape.